Fluid Power
Educational
Series

Design Concepts in Industrial Hydraulic Systems
(In the English Units)

Joji Parambath

Design Concepts in Industrial Hydraulic Systems
(In the English Units)

Copyright © 2021 Joji Parambath

All rights reserved

ISBN: 9798653933868

https://jojibooks.com

First Edition – 2020
Revised Edition - 2021

Disclaimer of Liability

The contents of this book have been checked for accuracy. Since deviations cannot be precluded entirely, we cannot guarantee full agreement. Only qualified personnel should be allowed to install and work hydraulic equipment. Qualified persons are defined as persons who are authorized to commission, to ground, and tag circuits, equipment, and systems following established safety practices and standards.

Dedicated to

my dearest friend Suresh

Table of Contents

PREFACE

The book describes the design aspects of hydraulic systems systematically. It highlights the essential parameters and specifications of hydraulic components. Some examples of designing typical hydraulic systems are also given in this book. The book uses the English system of units.

However, it may be noted that the book is intended for general educational purpose to illustrate the essential principles, techniques, and procedures. The optimum design of a hydraulic system depends on the exact operating and environmental conditions, amongst other factors.

Many other fluid power topics are given in other textbooks under the fluid power educational series by the same author. A list of all the textbooks is given at the end of this book (Page No. 163). Also, please see the details at https://jojibooks.com

Enjoy reading the book.
Your feedback is most welcome.

JOJI Parambath

Chapter 1 | Design Considerations

A hydraulic system must be designed to meet all the functional requirements of an application safely and efficiently. It is also essential to prepare the circuit diagram of the system using the correct ISO/ANSI/CETOP symbols, according to the norms in one's region. The system must provide the required performance level, withstand operational hazards, and ensure its life expectancy.

The design must also facilitate easy maintenance and the efficient removal of contaminants. Safety must be built into the system by incorporating interlocks, power-failure locks, and an emergency shutdown feature. It must also take into account of the speed of operation, the pressure and temperature ratings, the quality of components, the cost of downtime and component replacement, the sensitivity to contamination, and the environmental conditions. Further, the hydraulic system to be developed must be economical. It must avoid any overload, wear of system parts, overheating of the system fluid, over-sizing, and high cost.

General Design Principles
Industrial hydraulic systems are designed with correctly-sized components and conductors. The use of undersized components and conductors in a system can cause excessive pressure losses resulting from friction, and as a consequence, the operating cost increases significantly. In contrast, the use of oversized components and conductors can impose higher capital and installation costs.

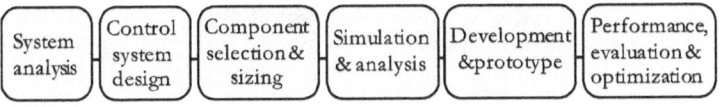

Figure 1.1 | Different stages in the design process

A simple and systematic approach is always a better way to design a hydraulic system. A typical approach consists of following a set of critical steps, as outlined in the flow diagram of Figure 1.1. These essential design steps are: (1) System analysis and drawing specifications, (2) Circuit/Control system design, (3) Component selection and sizing, (4) Software simulation and analysis, (5) Development of system prototype, and (7) System performance evaluation & optimization. The designer must take into account all working conditions specific to a given application. The following sections describe the essential steps for finding the correct sizes of components used in hydraulic systems.

System Analysis

The initial step in the design process of a hydraulic system is to define the exact requirements of the system. For understanding and defining the system requirements, it may be necessary to carry out detailed system analysis and develop operational specifications and system schematics.

It is essential to find the magnitude of each force (torque) and the type of motion required in the system. The requirements of the type, size, stroke, duty cycle, and the speed of actuators, and the mounting styles of components must be investigated.

The sequence of operations required in the application is to be clearly understood and detailed.

The type of the fluid medium to be used, the materials of construction of components, and the environment in which the system would be operated are other essential factors that must be considered while designing the system.

It is also necessary to determine the need for alternative control possibilities of the system, such as the use of fixed or variable displacement pumps, multiple pump combinations, accumulators, load sensing, and closed-loop controls.

It may be required to describe many factors, such as load characteristics, actuator characteristics, sensor characteristics, the extent of acceptable leakage, and fluid characteristics, while designing the system.

Factors, such as temperature, vibration, shock, exposure to outdoor weather, moisture, the possibility of chemical contact, initial costs, and maintenance costs, must also be considered while designing the system.

Many general requirements of the system, such as its robustness, compactness, quality, performance, efficiency, reliability, and safety, cannot be ignored.

An appropriate fluid cleanness level in the system, as per the relevant standards, needs to be determined.

It may also be necessary to construct the 'pressure Vs time' and 'flow Vs time' diagrams, a draft circuit, and the machine layout of the system.

Manufacturers' technical data must be consulted when configuring a solution to the design problem.

Circuit Design
Once the system requirements are established, system and performance specifications must be prepared.

The next step is to develop suitable control circuits for the system using the symbols as per the relevant standard to meet the system specifications.

Compare any alternative circuit solutions concerning the power transmission efficiency and the system cost.

Industrial hydraulic circuits may be spread over many sheets. The circuits may be developed and drawn in a simplified manner as far as possible and are developed using standards symbols. The breaking of a complex circuit into small parts and then linking these small parts are essential for the proper representation of the entire circuit. Figure 1.2 shows a typical format of a simplified industrial hydraulic circuit with a manifold assembly.

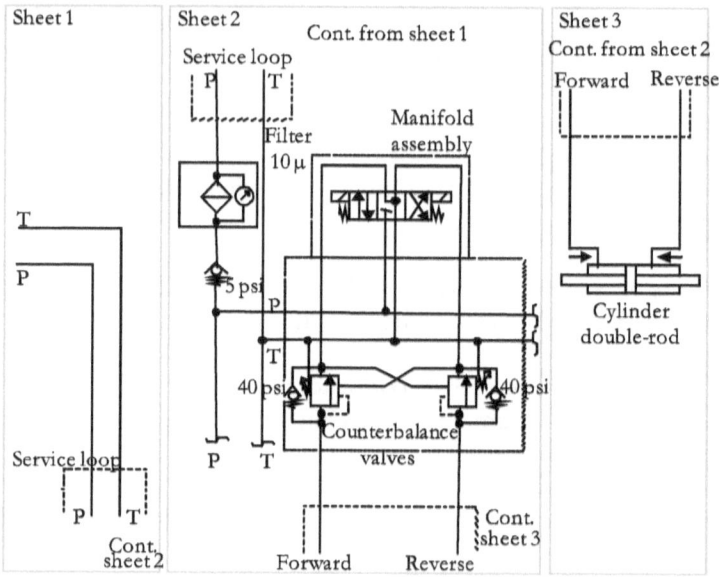

Figure 1.2 | The typical layout of a simplified industrial hydraulic circuit with a manifold assembly

Component Selection

The first step in the selection of hydraulic components for the system is to outline the key factors that affect the selection of the components. For example, the cycle time of work operations in an application is a critical factor when deciding the need for a double-acting cylinder. It is also necessary to specify the requirement of a safety block along with a hydraulic accumulator if used in the system.

Component sizing

Building the right hydraulic system for the specific application requirements is best achieved by first determining the parameters of the components of the system. The size of actuators, valves, pumps, conductors, and accessories to meet the performance specifications of the system must be determined. It may be noted that manufacturers offer hydraulic components with their sizes graded.

Simulation and Analysis

An appropriate software package can assist the designer in studying the interactions of the components of the system and analytically evaluating the performance of the system. The software usually assists in component-level modelling as well as the system-level modelling and assessing their steady-state and dynamic performances.

Development of Prototype

Next, a prototype of the system must be developed to analyze, evaluate, and optimize the actual performance characteristics of the system.

Performance Evaluation

The performance of the newly-developed system must meet the required specifications, especially concerning its power, torque/force, speed, and efficiency, under the specified operating conditions. Verify that the pressure and flow generated by the system meet the requirement specifications under all working conditions. Also, verify that all the cylinders and motors in the system have enough strength to drive the associated loads and have enough capacity to accommodate side loads if any. Ensure that the pressure losses, power losses, and heat generated in the system, under the worst operating conditions, are within limits. If any shortcomings are observed, the system must be modified for its optimum performance.

Chapter 2 | Review of Mechanics

Mass (m): Mass of a body is an attribute of the body that determines the effect of a force applied to it. In the English system, its unit is the slug. One slug is the mass of a body that accelerates at 32.2 ft/s² when a force of 32.2 pounds (lb) acts upon it.

Weight (W): Weight is related to the force of gravity (g) and is given by the relation: W = mg. In the English system, its unit is the pound (lb). One pound is the force required for accelerating one slug of mass at 32.2 ft/s².

Volume (V): The volume (V) of a liquid is a measure of how much three-dimensional space it occupies. In the English system, the unit of volume is cubic-foot, cubic-inch or gallon. 1 ft³ = 1728 in³

Density (ϱ): It is the physical property of a material that measures its quality of being dense. In the English system, its unit is slug/ft³.

Specific Weight (γ): The specific weight of an object is its weight (W) per unit volume (V). That is, γ = W/V. In the English system, its unit is pound/ft³. The specific weight of water works out to be 62.4 lb/ft³. Most of the hydraulic oils have specific weights in the range from 55 to 58 lb/ft³.

Specific Gravity (SG): The specific gravity of an object at a given temperature is the fraction of its density over the density of water at the same temperature. Specific gravity is a dimensionless parameter.

Force (F): A force is any influence capable of creating change in the state of a body. It may be either a push or a pull. In physics, a fundamental law states that 'force' is equal to 'mass' (m) multiplied by 'acceleration' (a), that is, F = ma.

Work (W): Work is said to be done by an object by the application of a force when the force acts to move the object. The work done (W) by the object can be expressed as the product of the force (F) exerted on the object and the distance (d) through which it moves. That is, $W = F \times d$. In the English system, its unit is inch-pound (in.lb) or foot-pound (ft.lb).

Power: It can be expressed in terms of the rate of doing work. That is, Power = Work/Time = $F \times d/t = F \times v$. In the English system, the unit of power is measured in terms of in.lb/s, in.lb/min, ft.lb/s, ft.lb/min or horsepower (HP).

Horse Power (hp): hp is calculated by using the relationship given below:

$$hp = \frac{F\ (lb)\ x\ v\ (ft/s)}{550}$$

Torque (T): It is defined as the twisting force of a rotating device about the axis of rotation through a pivot point. It can be calculated by multiplying the applied force (F) by the distance (r) from the pivot point to the peripheral point where the force acts. That is, $T = F \times r$. In the English system, its unit is (in.lb or ft.lb).

Torque – Power Relations: In the English system of units, if the torque (ft.lb or in.lb) and the angular speed ω [(rad/s) or N (rpm)] of the rotating device are known, then the theoretical power in Horsepower (hp) conveyed by the device can be calculated by the following relationship:

Power (hp)
$= T\ (ft.lb)\ x\ \omega\ (rad/s)\ /550$
$= T\ (ft.lb)\ x\ N\ (rpm)\ /\ 5252$
$= T\ (in.lb)\ x\ N\ (rpm)\ /\ 63025$

Energy: It is the capacity of an object to do work, and it is calculated by multiplying the force acting on the object and the distance through which the object moves. Its unit in the English system is the foot-pound.

Chapter 3 | Hydraulic Fundamentals

Pascal's Law

Pascal's law states that 'Pressure at any one point in a static fluid is the same in every direction' and 'the pressure exerted on a confined fluid is transmitted equally in all directions, acting with equal force on equal areas'.

Hydraulic Pressure

Figure 3.1 shows a cylinder chamber filled with a definite volume of fluid and a piston. Consider that force (F) is applied to the fluid through the piston. When the fluid is pushed, its pressure (P) goes up in direct proportion to the applied force and inverse proportion to the piston area (A). That is, 'pressure is the force acting per unit area'. i.e., $P = F/A$

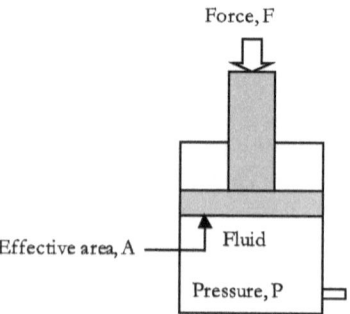

Figure 3.1 | A fluid-filled cylinder developing a definite amount of pressure with the application of a force

Pressure Units

In the US customary units, the quantity pressure has units of lb/ft^2 or lb/in^2. The unit of lb/in^2 is usually written as 'psi'. The bar to psi relationship is: 1 bar = 14.5 psi.

1 Pascal	$= 1 \ N/m^2$
1 bar	$= 10^5$ Pascal
1 Mega Pascal (MPa)	$= 10^6$ Pascal (10 bar)
1 bar	$= 0.1$ MPa

Pressure Levels in Hydraulic Systems

Standard operating pressures in industrial hydraulic systems range up to 5000 psi. High-pressure hydraulic systems operate in the range up to 10000 psi. Experimental extra high-pressure hydraulic systems operate in the range up to 50000 psi. Within the standard operating pressures, hydraulic systems can be classified as per the range given in Table 3.1:

Table 3.1 | Pressure levels in standard hydraulic systems

Standard pressure, Sub Division	Range (psi)
Low standard pressure	<1500
Medium standard pressure	1500 to 3000
High standard pressure	3000 to 5000

Hydraulic Force

When pressure (P) is applied to the area (A) of the piston of a cylinder, it develops a force (F). That is, 'force is equal to the applied pressure times the area'. i.e., $F(lb) = P(psi) \times A(in^2)$

Viscosity

The viscosity of a fluid can be measured in terms of its resistive movement when subjected to a force. Interestingly, there are two viewpoints of viscosity. One is the absolute viscosity (or dynamic viscosity), and the other one is the kinematic viscosity.

The absolute viscosity is the property that represents the resistive movement of different layers of fluid when subjected to an external force. The kinematic viscosity is the property that describes the difficulty with which the fluid moves under the force of gravity.

Absolute Viscosity (μ)

A thin plate A of surface area 'a' is located at a distance 'd' from a stationary reference plate B, as shown in Figure 3.2. The plate A is subjected to a force 'F' and moves with the velocity 'v'. For small values of v and d, the velocity gradient of the particles of the fluid layers tends to be a straight line with a slope v/d.

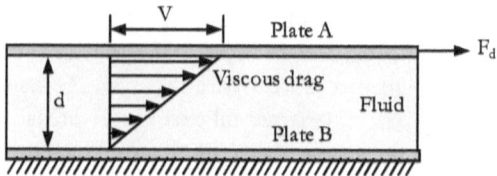

Figure 3.2 | Fluid velocity profile between two parallel plates

$$\text{Absolute viscosity, } \mu = \frac{(F/a)}{(v/d)}$$

Units of Absolute Viscosity
1 Poise = 1 dyne second per square centimeter (1 dyne.s/cm²)
1 centipoise (cP)   = 0.01 Poise
1 Pascal second (Pa.s)   = 1 N.s/m²
1 Poise   = 0.1 Pa.s

Kinematic viscosity (ν)
Kinematic viscosity is the measure of a fluid's resistance to flow under gravity. This measure at a given temperature is given by the absolute viscosity (μ) divided by the fluid density (ϱ).

$$\text{Kinematic viscosity, } \nu = \frac{\mu}{\varrho}$$

Units of Kinematic Viscosity
1 Stoke   = 1 cm²/s
1 Pascal second (Pa.s) = 1 N.s/m² = 0.02088543 lb.s/ft²
1 cSt = 0.00155 in²/s = 0.00001076391 ft²/s

Other Kinematic Viscosity Units
The measurements of viscosity frequently use the force of gravity to produce flow through a specially-designed capillary tube (viscometer) at the specified temperature. In addition to the basic units of measuring the kinematic viscosity, there are other units, such as Saybolt Universal Seconds, used for expressing the kinematic viscosity.

Saybolt Universal Seconds (SUS)

It is the time measured in seconds required for 60 ml of a sample of test fluid to pass through the calibrated orifice of a Saybolt Universal viscometer at the specified temperature, as covered by the empirical test method specified in the standard ASTM D88. The relationships between various units of kinematic viscosity (ν) and absolute viscosity (μ) are given below:

$$\nu \ (cSt) = \mu \ (cP) \ / \ SG$$
$$\nu \ (SSU) = 4.63 \ \mu \ (cP) \ / \ SG$$

Where SG = Specific gravity

A viscosity measure given in SSU can be converted to the corresponding viscosity in cSt by the following formula:

$$cSt = SSU / 4.635$$

Viscosity Classification Systems

In 1975, the International Standards Organization (ISO), in unison with American Society for Testing and Materials (ASTM) and many other standards organizations, settled upon an approach to establish a viscosity measurement method. It is known as the ISO Viscosity Grade (VG).

The ISO VG classification, as per the ISO standard 3448:1992, consists of a series of 20 different viscosity grades. Some of the Viscosity Grades are as follows: 2, 3, 5, 7, 10, 15, 22, 32, 46, 68, 100, 150, 220, 320, 460, 680, 1000, and 1500.

Each ISO VG number is the mid-point of the pertinent viscosity range expressed in centistokes (cSt) at 40°C. For example, a fluid with the grade of ISO VG 22 points to the viscosity range of 22 cSt ± 10% at 40°C.

The Effect of Variation in Pressure on Viscosity
An increase in the system pressure can cause an increase in the viscosity of the fluid. Figure 3.3(a) shows the distinct characteristics of viscosity versus pressure.

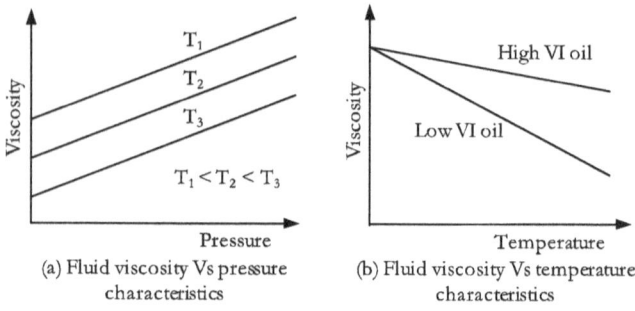

(a) Fluid viscosity Vs pressure characteristics

(b) Fluid viscosity Vs temperature characteristics

Figure 3.3 | Fluid viscosity Vs pressure characteristics

The Effect of Variation in Temperature on Viscosity
The viscosity of fluids can change appreciably with a change in their temperature. Fluids have higher viscosity when they are cold and lower viscosity when they are hot. Figure 3.3(b) shows the distinct characteristics of viscosity versus temperature.

Viscosity Index (VI)
The change in viscosity with temperature is measured with an arbitrary measure called Viscosity Index (VI). Fluid having a low VI exhibits a significant change in viscosity with temperature change. A High VI fluid has relatively stable viscosity, which does not change appreciably with temperature change.

Typical values of VIs for the petroleum fluids range from 90 to 105, and those for the synthetic fluids range from 80 to over 400. Hydraulic fluids can typically be selected with VI values in the range from 90 to 110.

Flow Rate, (Q)

The volumetric flow rate, in general, is a measure of the volume of the fluid passing a given cross-sectional area per unit of time. It is measured in cubic foot per second (ft^3/s) or gallons per minute (gpm). The flow rate in a hydraulic system is related to the speed requirements of actuators in the system.

Flow Velocity (v)

Specific flow velocities are found to be most favourable in certain distinctive parts of hydraulic systems, such as suction lines, pressure lines, and return lines. The recommended flow velocities for initial pipe sizing are given in Page 66 and 67, Chapter 13.

Flow Rate Vs Velocity of Flow

A fluid is pumped at a constant flow rate through the system pipeline with two sections having cross-sectional areas A1 and A2 (A1 > A2), as shown in Figure 3.4. When the fluid passes through the pipeline at its narrow part, the fluid speed goes up.

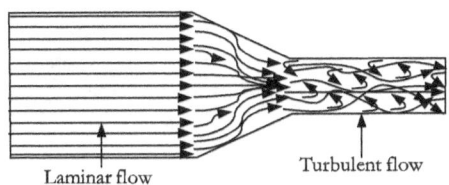

Laminar flow   Turbulent flow

Figure 3.4 | A pipe with a larger section and a smaller section

It is obvious that at a constant flow rate (Q), the velocity (v) of the fluid increases, as it passes through the narrow section of the pipe, following the relation given below. In the English units,

$$\text{Flow rate } (Q, ft^3/s) = \text{Area } (A, ft^2) \times \text{Velocity } (v, ft/s)$$

Laminar & Turbulent Flows

The flow of fluid through a hydraulic system can be laminar or turbulent. The laminar flow is a smooth flow through the system. The turbulent flow results in a significant increase in the friction, flow resistance, pressure drop, and energy loss in the system, as compared to the case of a laminar flow.

An essential objective of any hydraulic system is to avoid the turbulent flow and maintain laminar flow through the system. This goal can be achieved by: (1) adequately sizing the system components and the fluid conductors, (2) avoiding abrupt changes in the system pipe cross-section and the sharp turns of the piping, and (3) limiting the velocity of the system fluid.

Osborn Reynolds conducted a series of pioneering works investigating the nature of fluids and determining the governing condition regarding the transition of fluid flow from laminar state to the turbulent state.

Reynolds Number (R_e)

Reynolds number is an important design parameter in hydraulic systems. Osborn Reynolds defined a dimensionless parameter, in respect of fluid flow through pipelines, to characterize the transition of fluid flow from the laminar state to the turbulent state. The ratio of the inertial forces to the viscous forces of the flow is the Reynolds number. The equation below gives this number.

$$R_e = \frac{v\,D\,\varrho}{\mu} = \frac{v\,D}{\upsilon}$$

Where,

v	= Fluid velocity, ft/s	
D	= Internal diameter of the pipe, ft	
ϱ	= Fluid density, slug/ft³	
μ	= Absolute viscosity of the fluid, lb·s/ft²	
υ (nu)	= Kinematic viscosity, ft²/s	

Using the kinematic viscosity in cSt, the Reynolds number is given by:

$$Re = \frac{7740 \times v(ft/s) \times d(in)}{\nu\,(cSt)}$$

The number 2000 is the decisive value of the Reynolds number for marking the borderline between the laminar flow and the turbulent flow. Experiments have demonstrated that for Reynolds number below 2000, the flow is found to be laminar, and above 4000, the flow is found to be turbulent. The Reynolds number regime between 2000 and 4000 can be considered as the critical zone. The flow of the fluid in the critical zone is usually assumed to be turbulent for all practical purposes.

Compressibility and Bulk Modulus of Hydraulic Fluids

Hydraulic fluids are commonly assumed to be incompressible. In practice, however, they are not entirely stiff, but they show some degree of compressibility when subjected to high pressures. The effect of compression shows up as a loss of power.

The compressibility of the fluid is the degree to which the fluid undergoes a reduction in its volume, as the pressure exerted on the fluid increases. The compressibility of the fluid is the reciprocal of its bulk modulus. The English unit of the bulk modulus is psi. The following equation gives the bulk modulus.

$$B = - \Delta P / (\Delta V / V)$$

Where,
B   =Bulk modulus, in psi
ΔP =Differential change in the pressure, in psi
ΔV =Differential change in the fluid volume, in in^3
V   =The original volume of the fluid, in in^3

The bulk modulus of a newly purchased hydrocarbon-based fluid is typically about 250000 psi.

Chapter 4 | Hydraulic Fluids

The fluid in a hydraulic system essentially transmits power from one point to another. It is prepared from a base stock and additives.

Categories of Hydraulic Fluids

As the modern hydraulic systems require the high-performance hydraulic fluids to meet the stringent requirements of these systems, manufacturers prepare varieties of hydraulic fluids apart from the mostly used mineral-based fluids. Four basic types of hydraulic fluids are evolved over a period. Two categories of hydraulic fluids are explained below:

Mineral-based Fluids (Petroleum-based Fluids)

The petroleum-based fluids possess most of the desired properties required for hydraulic power transmission. The main advantages of using the petroleum-based fluids in hydraulic systems are their inherent lubricating and corrosion-inhibiting properties. They are the low-cost fluids available in a broad range of viscosities. Also, their properties can be improved, and their service life can be extended by blending them with suitable additives.

However, the major drawbacks of the mineral-based fluids are that they are flammable and can become explosive when they are subjected to high pressures or temperatures, or both. They are also toxic and not very much bio-degradable. In general, petroleum-based fluids are used in systems where the possibility of fire hazards is comparatively little.

Fire-resistant Fluids

There is a growing demand for effective fire-resistant fluids for the high-temperature or hazardous hydraulic applications, especially in mines and steelworks. There have been many fluids developed for applications sensitive to the fire hazard. Two basic types of fire-resistant fluids are: (1) High-water-based-fluids (HWBF) and (2) Synthetic fluids.

Classification of Fire-resistant Hydraulic Fluids

ISO 6743-4 /CETOP RP 77H divides the fire-resistant hydraulic fluids into the groups HFA, HFB, HFC, and HFD. The following lines give a few examples of these types of fluids:

- **HFA:** Oil-in-water emulsions with a combustible proportion of 20% maximum.
- **HFB:** Water-in-oil emulsions with a combustible proportion of 60% maximum. The HFB fluids are used extensively in mining machines and some steel industries.
- **HFC:** Water glycol solutions with a water proportion of at least 35%. The HFC fluids are used almost exclusively in mining machines with open hydrostatic circuits.
- **HFD:** Water-free fluids on a synthetic base.

Fluid Cleanness Standards

Many national and international organizations, including ISO, ASTM, and SAE, have developed standards for specifying the 'particle size classification' and the 'contamination concentration levels' of hydraulic fluids.

ISO 11171:2010 specifies three-dimensional sizes of particles (i.e., 4, 6, and 14 microns), as specified in the ISO 4402 standard, for representing the concentration levels of fine as well as coarse particles.

As per ISO 4406:1999 standard, the cleanness level of a given sample of fluid can be defined by the three-dimensional range code representation, such as 18/16/14, based on the numbers of particles of sizes greater than 4, 6, and 14 microns respectively present in one ml of the sample fluid. Table 4.1 provides the range of particles per ml for each of the range codes from 1 to 30.

Table 4.1 | Contamination code rating system as per ISO 4406

Range Code	Number of particles	
	>	<=
1	0	0.02
2	0.02	0.04
3	0.04	0.08
4	0.08	0.15
5	0.15	0.3
6	0.3	0.6
7	0.6	1.3
8	1.3	2.5
9	2.5	5
10	5	10
11	10	20
12	20	40
13	40	80
14	80	160
15	160	320
16	320	640
17	640	1,300
18	1,300	2,500
19	2,500	5,000
20	5,000	10,000
21	10,000	20,000
22	20,000	40,000
23	40,000	80,000
24	80,000	160,000
25	160,000	320,000
26	320,000	640,000
27	640,000	1,300,000
28	1,300,000	2,500,000
29	2,500,000	5,000,000
30	5,000,000	10,000,000

For example, consider the measured cleanliness level 18/16/14 of the hydraulic fluid. The three numbers 18, 16, and 14 represent the range codes for the numbers of particles of sizes greater than 4, 6, and 14 microns respectively, present in one ml of the fluid. It can be seen from the Table that corresponds to the range code 18, the number of particles of size greater than 4 microns present in the fluid lies in the range from 1301 to 2500. Similarly, the numbers of particles of sizes greater than 6 microns and 14 microns corresponding to the range codes 16 and 14, respectively in one ml of the fluid can be found out from the Table.

Typical Cleanliness Level Targets for Components

Hydraulic equipment manufacturers, fluid suppliers, and fluid power associations have established target fluid cleanliness levels applicable to the general types of hydraulic components. In Table 4.2, a few components and their typical target cleanliness levels, using the ISO range codes are given in the most generalized way.

Table 4.2 | Typical cleanness levels, using petroleum oil, for hydraulic components

Components	System Pressure Level		
	<2000 psi	2000–3000 psi	>3000 psi
Vane pumps, fixed	20/18/15	19/17/14	18/16/13
Vane pumps, variable	18/16/14	17/15/13	--
Piston pumps, fixed	19/17/15	18/16/14	17/15/13
Piston pumps, variable	18/16/14	17/15/13	16/14/12
Directional valves	20/18/15	20/18/15	19/17/14
Proportional valves	17/15/12	17/15/12	15/13/11
Servo valves	16/14/11	16/14/11	15/13/10
Pressure/Flow control valves	19/17/14	19/17/14	19/17/14
Cylinders	20/18/15	20/18/15	20/18/15
Vane motors	20/18/15	19/17/14	18/16/13
Axial piston motors	19/17/14	18/16/13	17/15/12
Radial piston motors	20/18/14	19/17/13	18/16/13

NAS Code

The National Aerospace Standard (NAS) 1638 coding system defines the maximum numbers permitted of 100 ml volume at various size intervals (for aircraft systems)

Table 4.3 | Maximum contamination limits (per 100 ml)

ISO 4406 Equivalent	NAS code	5-15 µm	15-25 µm	25-50 µm	50-100 µm	> 100 µm
—	00	125	22	4	1	0
—	0	250	44	8	2	0
12/10/7	1	500	89	16	3	1
13/11/8	2	1000	178	32	6	1
14/12/9	3	2000	356	63	11	2
15/13/10	4	4000	712	126	22	4
16/14/11	5	8000	1425	253	45	8
17/15/12	6	16000	2850	506	90	16
18/16/13	7	32000	5700	1012	190	32
19/17/14	8	64000	11400	2025	360	64
20/18/15	9	128000	22800	4050	720	128
21/19/16	10	256000	45600	8100	1440	256
22/20/17	11	512000	91200	16200	2880	512
23/21/18	12	1024000	182400	32400	5760	1020

SAE Aerospace Standard AS4059

The SAE aerospace standard AS4059, for specifying particulate contamination in hydraulic fluids in different classes, was developed in 1988 as a replacement to the NAS 1638 standard. The details of the standard AS4059 are given in Annexure 7.

Chapter 5 | Hydraulic Filters

The degree of fluid cleanness achieved by a hydraulic system fluid can be linked to the performance of the filter elements used in the system. These filter elements are rated based on their ability to separate the contaminants of particular sizes from the system fluid, under the specific test conditions. Amongst others, the most critical parameters are the mesh number, Beta (ß) Ratio, filter efficiency, and the micron ratings. The following sections describe these two parameters.

Mesh Number/Sieve Number
The mesh size or fineness of a wire-mesh filter can be expressed in terms of its mesh number or sieve number. It is the number of openings from the center of any wire of the wire mesh to the center of the parallel wire one inch away. For example, a 2-mesh screen has two openings across one linear inch of the screen, and a 100-mesh screen has 100 openings.

Beta Ratio
The Beta (ß) ratio of the filter is also known as the filtration ratio. It signifies the effectiveness of the filter element in removing the contaminants from the associated hydraulic system. Figure 5.1 shows the schematic diagram of the partial test setup for measuring the Beta ratio of the filter.

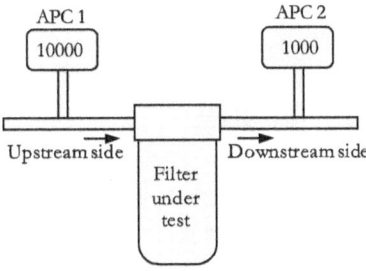

Figure 5.1 | Test setup for measuring the Beta ratio of the test filter

In this test, it is necessary to pump a calibrated fluid with a homogeneous amount of contaminants, through the test filter. Then, count the numbers of particles equal to or larger than the specified size, upstream and downstream of the test filter, using the automatic particle counters (APCs). The Beta ratio can then be determined by using the formula given below:

$$\text{Beta ratio}_{x(c)} = \frac{\text{Particle count in the upstream fluid}}{\text{Particle count in the downstream fluid}}$$

Where the subscript 'x' stands for the specified particle size and subscript '(c)' refers to the 'certified' calibration. That is, the subscript, (c), indicates the test adherence to the new ISO 16889 standard.

Filter Efficiency
This efficiency, in terms of percentage, can be found by the following equation:

$$\text{Efficiency}_{(x)} = \left(1 - \frac{1}{\beta}\right) \times 100$$

Beta Ratio and Filter Efficiency
Table 5.1 gives the values of the Beta ratio and the corresponding efficiency of a hydraulic filter for various combinations of upstream and downstream particles.

Table 5.1 | Beta ratios and the corresponding filter efficiencies

Upstream particles ($\geq$ x μm)	Downstream particles ($\geq$ x μm)	Beta (β_x) ratio	Efficiency$_{(x)}$
1,00,000	50,000	2	50.0%
1,00,000	5,000	20	95.0%
1,00,000	1,333	75	98.7%
1,00,000	1,000	100	99.0%
1,00,000	500	200	99.5%
1,00,000	100	1000	99.9%

Micron (μm) Rating, Filter

There are two most popular micron ratings for the hydraulic filters. They are: (1) absolute micron rating and (2) nominal micron rating.

Absolute Micron Rating of a filter is the smallest size of particles it can capture in excess of 98.6% on the first pass through it. It indicates the size of the largest opening in the filter.

Nominal Micron Rating of a filter is the smallest micron size of particles it can capture in a specified quantity, in the range from 50% to 95% on the first pass through it.

Differential Pressure (ΔP), Filter

The differential pressure (ΔP) across the filter, while in operation in a hydraulic system, indicates the difference between its inlet and outlet pressures. It points to a permanent loss of pressure in the system.

Particle Capture Efficiency, Filter

The particle capture efficiency (dirt holding capacity) of a filter indicates the quantity of the solid dirt that its filter element can hold before it has to be replaced. It is the weight of the specified artificial contaminant (ISO medium test dust) that must be added to the fluid, upstream of the filter, to produce a given pressure differential across the filter under the specified test conditions. This efficiency indicates the service life of the filter and hence is an essential parameter of the filter.

Burst Pressure, Filter

The burst pressure of a filter is also known as the 'collapse pressure'. It is the minimum inside-out pressure differential that a spin-on hydraulic filter can withstand without the outward structural or media failure.

Chapter 6 | Hydraulic Reservoirs

A hydraulic system requires a sufficient amount of high-quality fluid at all times for its efficient operation. A power pack, as shown in Figure 6.1, is a unit that supplies the required fluid to the system. It is a compact, portable, and custom-designed or pre-engineered assembly consisting of essential and optional components. The essential components are a fluid-filled reservoir, close-coupled pump-motor unit, pressure relief valve, and pressure gauge, and the optional components include a heat exchanger, temperature controller, directional control valves, filters, etc. It also consists of necessary instrumentation, and other accessories, such as accumulators, hoses, and quick-disconnect couplings. The modern way of configuring a power pack is from standardized sub-assemblies.

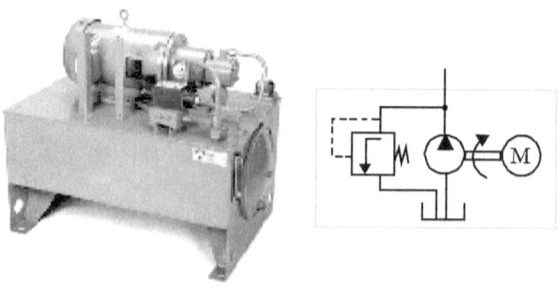

Figure 6.1 | Hydraulic power pack

Constructional Features of hydraulic Reservoirs
A well-designed reservoir should be completely enclosed and self-contained. Figure 6.2 shows the cross-sectional view of a reservoir. It is provided with the following essential and optional parts: (1) Tank with top plate, (2) Baffle plate, (3) Suction line, (4) Return line, (5) Filler-cum-breather, (6) Drain plug, (7) Strainer, (8) Fluid level indicator, (9) Pressure gauge, (10) Removable cover, (11) Diffuser, (12) Magnetic tank cleaner, and (13) Heater. Hydraulic reservoirs are modelled in vertical and horizontal designs.

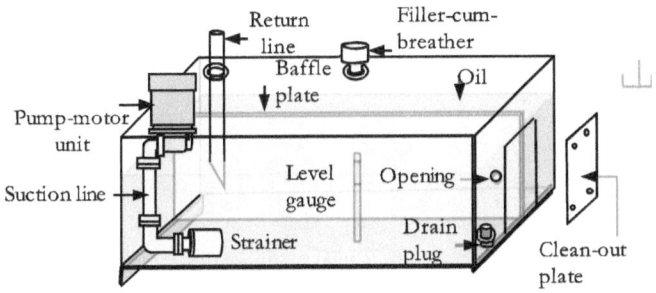

Figure 6.2 | Hydraulic reservoir

The thickness of the tank plate can range from 14 to 7 gauge.

Sizing of Hydraulic Reservoir

The size of a hydraulic reservoir depends on the application concerned. For a hydraulic system, where the mineral fluids are used, and medium-to-high frequency demands are expected, a reservoir with a capacity (gallons) of three to five times the volume flow rate (gpm) of the system fluid is adequate. That is,

Reservoir size, gallons = (3 to 5) x pump flow rate, gpm

However, the recommended size of a reservoir for a hydraulic system should be larger than normal under certain exceptional situations. That is, a larger reservoir should be used if it is liable to be exposed to high ambient temperatures or if some special fluids are intended to be used in the system.

For example, a reservoir size of five to eight times the pump flow rate per minute is recommended for a hydraulic system with a water-glycol or Polyol ester fluid medium. Further, the reservoir needs to be larger if the associated system has accumulators. However, remember, with the use of a heat exchanger in a system, the reservoir size could be as small as that required to accommodate the fluid corresponding to the pump flow rate in one minute.

A holistic approach to finding the most appropriate size of a reservoir in a hydraulic system is to consider the heat balance feasible in the system. Towards this end, it is necessary to find the amount of heat generated in the system as a result of losses, and then to determine the heat-dissipating area of the reservoir and how much heat can be dissipated from that area. From these calculations, the optimum size of the reservoir and the size of the heat exchanger can be determined. Heat-dissipation factor, which is an essential consideration in deciding the reservoir capacity, is explained in the next section.

Heat Radiation from a Reservoir

The fluid is at room temperature when the machine has been started. As time passes, the fluid temperature will start rising until a levelling-off temperature is reached. Under this condition, the heat is radiated as fast as it is produced. If the levelling-off temperature is acceptable, then the system will take care of itself; otherwise, a heat exchanger is to be added to bring the temperature to an acceptable level. Selecting the size of a heat exchanger is not an exact science as there are too many unknown factors which cannot be predicted accurately. It should always be selected oversize.

Heat Load of a New Reservoir

The heat load of a reservoir that is being designed can be approximately derived from the input power to the system as specified at the motor nameplate or engine rating, as a guide. The losses can be approximately taken as 15% of the input power for the pump and hydraulic motor and 20% for the valves. A system with a pump and a cylinder, the system losses would be about 30% of the input power. The associated reservoir would dissipate a part of the heat, and the remainder could be handled with a heat exchanger.

Heat Load of an Existing Reservoir

The first step to calculate the heat loss of an existing reservoir is to measure the tank temperature at the startup. Then, measure the temperature difference (ΔT) after the system has been in operation for a specified duration (Δt). Then the heat load can be calculated as:

$$\text{Heat load (hp)} = \frac{V \text{ (gallons) x Cp (btu/lb) x } \varrho \text{ (lb/ft}^3\text{) x } \Delta T \text{ (}^\circ\text{F)}}{317.3 \text{ x } \Delta t \text{ (min)}}$$

Where,
Cp is the specific heat of the fluid in btu/lb ($^\circ$F)
ϱ is the density of the fluid in lb/ft^3
1 hp = 2544 btu/hour

Heat dissipation by Hydraulic Reservoirs

Assume that a reservoir with a heat-dissipating surface area 'A' dissipates heat 'H' through conduction and radiation. The base of the reservoir can be excluded from the heat-dissipating surface if it is not elevated by at least 6 inches from the ground. Let 'ΔT' is the temperature difference between the reservoir walls and the ambient air. The formulae for the heat dissipated by the reservoir in the English system of units are given by:

The general formula:
$$H = k \text{ x } \Delta T \text{ x } A, \text{ where k is a constant}$$
In the English system:
$$H \text{ (hp)} = 0.001 \text{ x } \Delta T \text{ (}^0\text{F) x } A \text{ (ft}^2\text{)}$$

Heat Exchangers

Heat is usually generated in a hydraulic system due to its inefficiency or its poor design or both. Devices used in the hydraulic system, such as the flow control valves, sequence valves, pressure reducing valves, and undersized directional control valves can contribute to the development of heat in the system. If the cooling effect from the reservoir is insufficient, a heat exchanger (or cooler) must be fitted to increase the heat dissipation rate of the system. However, the heat exchangers are expensive, and the maintenance of them can run high. Two main types of heat exchangers are used in hydraulic systems. They are: (1) air-cooled heat exchangers and (2) water-cooled heat exchangers.

Noise in Hydraulic Systems

Hydraulic systems can be a source of great noise. The sound is produced when a sound source makes the air medium nearest to it in vibratory motion and makes it move in a wave pattern with a particular frequency and intensity. The frequency is measured in hertz (Hz). The intensity of the sound is measured in terms of 'decibel' (dB). The frequency of the audible sound for human ears is in the range between 20 Hz to 20,000 Hz. Human beings are very much sensitive to the sound in the frequency range from 1 kHz to 4 kHz, rather than to very-low-frequency or high-frequency sounds. For this reason, the sound meter is usually fitted with a special filter, such as 'A-weighting' filter, whose response to frequency is similar to that of the human ear. If an A-weighting filter is used, the sound pressure level is given in dB (A).

Sources of Noise in Hydraulic Systems

There are many sources of noise in hydraulic systems. In general, higher noise levels in hydraulic systems are produced by the piston pumps and the relief valves. The sources of noise are categorized as: (1) structure-borne noise, (2) fluid-borne noise, and (3) airborne noise.

Noise Reduction Techniques

It is necessary to design hydraulic systems, especially the power units with appropriate noise reduction techniques to reduce the damaging effects of noise. In general, the noise in the machine can be controlled by: (1) using quieter work processes, (2) enclosing the machine to reduce the noise at source, and (3) using sound-absorbing materials in the machine to prevent the spread of the noise. The noise reduction can be achieved by isolating the motor-pump unit of the machine from its base, isolating the structural elements of the power unit that could intensify the sound, and using the hoses and the tubing correctly. An integrated motor-pump unit has very low sound levels. Many other factors, such as the mounting, tank style, and plant layout, affect the noise levels.

Chapter 7 | Hydraulic Pumps

The positive-displacement pumps come in many different varieties, sizes, flow rates, and power ratings. The cross-sectional view of a pump is given in Figure 7.1. Some important pump parameters are described below.

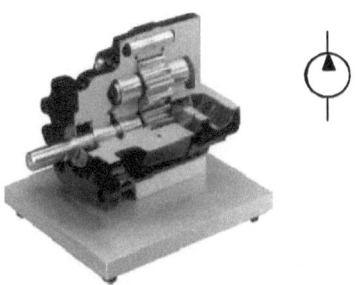

Figure 7.1 | Hydraulic pump

Pressure Rating, Pump: It is the pressure that overcomes all resistances in the system, which includes both useful work and losses. If an application involves simple or moderate work, a low to medium pressure pump would be the most suitable for the application. On the other hand, if an application requires substantial work, as in large construction equipment, a high-pressure pump would be the most appropriate.

Volumetric Displacement (V_D), Pump: It is the volume of the fluid that is carried by the pump in one revolution of its driveshaft. It is expressed in cubic-inch per revolution (in^3/rev), cubic-centimeter per revolution (cc/rev), or other similar units.

Theoretical Flow Rate (Q_T): It is the volume of the fluid displaced by the pump, at its inlet, per unit of time. It can be determined by the product of the volumetric displacement of the pump and the speed of the pump's driveshaft. In the English system of units, the theoretical flow rate is measured in cubic-inch per minute (in^3/min) or gpm.

The mathematical equations for the theoretical flow rate (Q_T) of the pump in the English system of units are as follows:

$$Q_T \text{ (in}^3/\text{min)} = V_D \text{ (in}^3/\text{rev)} \times N \text{ (rpm)}$$

[Note: As 1 gallon = 231 in³, Q_T (gpm) = Q_T (in³/min) / 231]

Pump Slippage (Q_s): It represents the internal leakage of the fluid in the pump from its discharge port to its suction port. The internal leakage is due to some unavoidable small clearance that exists between the internal parts of the pump. The slippage is a function of the pump speed, the differential pressure across the pump, the degree of wear of its interior surfaces, and the viscosity of the fluid passing through the pump. Any increase in the slippage leads to lesser efficiency of the pump.

Actual Flow Rate (Q_A): It is the actual fluid volume discharged by the pump per unit of time. It is given by the theoretical flow rate minus the pump slippage. That is,

Actual flow rate = Theoretical flow rate − Slippage

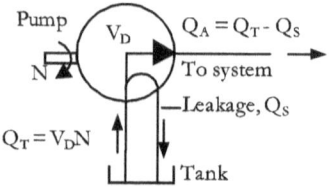

Figure 7.2 | A symbolic representation of a hydraulic pump showing its flow rate parameters

In the English system of units, the actual flow rate is measured in cubic-inch per minute (in³/min) or gpm. Figure 7.2 depicts the relationship between the theoretical and actual flow rates in the pump drawing fluid from the tank and delivering to the system.

Example 7.1: What is the theoretical flow rate of a fixed-displacement hydraulic pump with a volumetric displacement of 0.0046 ft³/rev operating at 2000 rpm?

Solution

Volumetric displacement, V_D = 0.0046 ft³/rev
Pump speed, N = 2000 rpm

Theoretical flow rate, Q_T = V_D x N
= 0.0046x2000
= 9.2 ft³/min

Actual Torque (T_A): It is the actual torque delivered to the pump by its prime mover and is given by:

$$T_A(\text{in.lb}) = \frac{\text{Actual power delivered to the pump (hp) x 63025}}{N \text{ (rpm)}}$$

$$T_A(\text{ft.lb}) = \frac{\text{Actual power delivered to the pump (hp) x 5252}}{N \text{ (rpm)}}$$

$$T_A(\text{ft.lb}) = \frac{\text{Actual power delivered to the pump (hp) x 550}}{n \text{ (rps)}}$$

Theoretical Torque (T_T): The theoretical Torque of the pump is a function of the volumetric displacement of the pump and the system pressure. It is equal to the actual torque minus the torque losses on account of the friction in the pump.

$$\text{Theoretical Torque, } T_T(\text{in.lb}) = \frac{V_D(\text{in}^3/\text{rev}) \text{ x P (psi)}}{2\Pi}$$

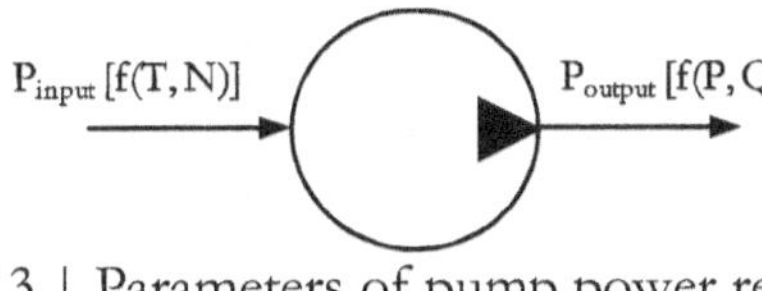

P_{input} [f(T,N)] P_{output} [f(P,Q)]

Figure 7.3 | Parameters of pump power relationships

Pump Input Power: It is the power delivered to the pump by its prime mover. Figure 7.3 shows the parameters of pump power relationships. The speed of the driveshaft and torque imparted by the motor determine the input power to the pump.

$$\text{Pump input power (hp)} = \frac{T_A(\text{in.lb}) \times N(\text{rpm})}{63025}$$

$$\text{Pump input power (hp)} = \frac{T_A(\text{ft.lb}) \times N(\text{rpm})}{5252}$$

$$\text{Pump input power (hp)} = \frac{T_A(\text{ft.lb}) \times n(\text{rps})}{550}$$

Pump Output Power: It is the power delivered by the pump. The pressure and the actual flow rate of the pump determine the output power.

$$\text{Pump output power (hp)} = \frac{P(\text{psi}) \times Q_A(\text{gpm})}{1714}$$

Example 7.2: A hydraulic pump delivers 10.6 gpm at 2176 psi for carrying out a work operation. Calculate the hydraulic power developed by it.

Solution

With usual notations,

Flow, Q_A = 10.6 gpm
Pressure, P = 2176 psi

$$\begin{aligned}
\text{Output power, } P_{out} &= P\ (\text{psi}) \times Q_A\ (\text{gpm})\ /\ 1714 \\
&= 2176 \times 10.6/1714 \\
&= 13.46 \text{ hp}
\end{aligned}$$

Efficiencies of Hydraulic Pumps

In a practical hydraulic pump, both fluid leakage and frictional loss take place and, therefore, their efficiency is always less than 100%. Two types of efficiencies are identified to account for the two types of losses in the pump. They are: (1) Volumetric efficiency, and (2) Mechanical efficiency.

Volumetric Efficiency (η_v)

It represents the ratio of the actual pump flow rate at a given pressure and the theoretical flow rate as determined by the geometric displacement of the pump, assuming no frictional losses. It indicates the extent of leakage that takes place within the pump. It is given by:

$$\text{Volumetric Efficiency} \left(\eta_v \right) = \frac{\text{Actual flow rate}}{\text{Theoretical flow rate}} = \frac{Q_A}{Q_T}$$

Mechanical Efficiency (η_m)

It represents the ratio of the power delivered by the pump to the power delivered to the pump, assuming no leakage in the pump. It indicates the amount of energy losses that take place in the pump due to friction. It is given by:

$$\text{Mechanical Efficiency} \left(\eta_m \right) = \frac{\begin{array}{c}\text{Pump output power,}\\\text{assuming no leakage}\end{array}}{\begin{array}{c}\text{Actual power delivered}\\\text{to the pump}\end{array}}$$

$$\text{Mechanical Efficiency} \left(\eta_m \right) = \frac{P_x \, Q_T}{T_A \times N}$$

The mechanical efficiency of the pump can also be calculated in terms of its torque units. That is,

$$\text{Mechanical Efficiency} \left(\eta_m \right) = \frac{T_T}{T_A}$$

Overall Efficiency (η_o)

It is the ratio of the actual power delivered by the pump to the actual power delivered to the pump. It is given by:

$$\text{Overall Efficiency } (\eta_o) = \frac{\text{Actual power delivered by the pump}}{\text{Actual power delivered to the pump}}$$

$$\text{Overall Efficiency } (\eta_o) = \frac{P \times Q_A}{T_A \times N}$$

The overall efficiency of the pump is also given by the product of its volumetric efficiency (η_v) and its mechanical efficiency (η_m). That is,

$$\eta_o = \eta_v \times \eta_m$$

Summary of Relations for Hydraulic Pumps

Figure 7.4 gives a summary of the essential relations of hydraulic pumps.

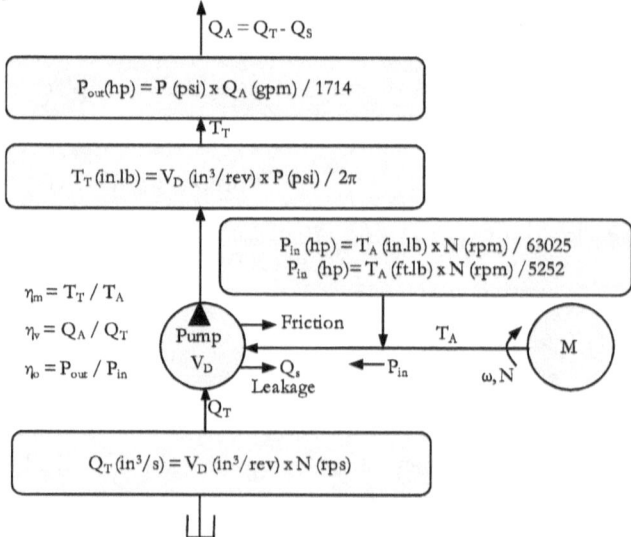

Figure 7.4 | Summary of relations for hydraulic pumps
Appendix 1 presents a typical pump data.

Example 7.3
A hydraulic pump with a displacement of 6.35 in³/rev delivers 25.4 gpm operating at a pressure of 1450 psi when driven by a prime mover with a torque of 1593 in.lb at 1000 rpm. Find out: (1) the volumetric efficiency, (2) the mechanical efficiency, and (3) the overall efficiency, of the pump.

Solution

Pump displacement, V_D	= 6.35 in³/rev
Actual flow rate, Q_A	= 25.4 gpm
Speed, N	= 1000 rpm
Pressure, P	= 1450 psi
Actual torque, T_A	= 1593 in.lb

Theoretical flow rate, $Q_T = V_D(\text{in}^3/\text{rev}) \times N$ (rpm)
$$= 6.35 \times 1000 = 6350 \text{ in}^3/\text{min}$$
$$= 6350/231 \text{ gpm} = 27.49 \text{ gpm}$$

Volumetric efficiency, $\eta_v = (Q_A/Q_T) \times 100\%$
$$= (25.4/27.49) \times 100\% = 92\%$$

Input power P_{in} = $T_A(\text{in.lb}) \times N(\text{rpm})/63025$
$$= 1593 \times 1000/63025 = 25.3 \text{ hp}$$

Output power, P_{out} = $P(\text{psi}) \times Q_T(\text{gpm})/1714$
$$= 1450 \times 27.49/1714 = 23.25 \text{ hp}$$

Mechanical efficiency, $\eta_m = (P_{out}/P_{in}) \times 100\%$
$$= (23.25/25.3) \times 100\% = 92\%$$

Overall efficiency, $\eta_o = \eta_v \times \eta_m = 0.92 \times 0.92$
$$= 0.85 = 85\%$$

Chapter 8 | Hydraulic Cylinders

Some essential parameters concerned with the operation and applications of hydraulic cylinders are its bore diameter, piston-rod diameter, force (thrust and pull), stroke length, speed, and piston-rod buckling. The following sections describe these terms:

Maximum operating pressure (P): It is the pressure that overcomes all resistances in the system, which includes both useful work and losses. Alternatively, it is the maximum working pressure that the cylinder can sustain without adverse consequence.

Bore Diameter (D): It refers to the diameter at the bore of the cylinder (See Figure 8.1). It can be used to calculate the bore area of the cylinder. It is also equal to the piston diameter, in a close-fitting hydraulic cylinder.

Piston-rod Diameter (d): It refers to the diameter of the piston-rod of the cylinder (See Figure 8.1).

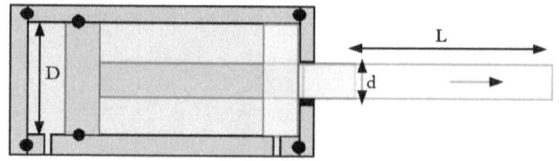

Figure 8.1 | Cylinder parameters

Stroke Length (L): It is the distance through which the piston-and-piston-rod assembly moves through the cylinder.

Maximum Stroke Length: It is the maximum linear movement that a cylinder can produce. For standard models of double-acting cylinders, the maximum stroke lengths can be up to 6.5 ft, and for special designs, the stroke lengths can be up to 20 ft.

Cylinder Thrust/Pull (F): The theoretical thrust (F) during the forward stroke or pull (F) during the return stroke of the cylinder can be determined by multiplying the effective area of the piston by the working pressure (P) to which it is subjected, according to Pascal's law.

The active area (A_{ext}), considered for the calculation of the cylinder thrust, is the full area (A_p) of the cylinder bore and is given by ($\pi.D^2/4$). The parameter 'D' denotes the piston (bore) diameter.

Further, the active area (A_{ret}), considered for the calculation of the cylinder pull, is the area (A_p) of the bore minus the piston-rod area (A_r), and it is given by the expression [$\pi. (D^2 - d^2)/4$]. The parameter 'd' denotes the piston-rod diameter.

The theoretical thrust and the theoretical pull are given by:

$$\text{Thrust, F (lb)} = P \text{ (psi) x } A_{ext} \text{ (in}^2)$$
$$\text{Pull, F (lb)} = P \text{ (psi) x } A_{ret} \text{ (in}^2)$$

Where,

A_p is the piston area
A_r is the piston-rod area
A_{ext} is the active area during extension: ($A_{ext} = A_p$)
A_{ret} is the active area during retraction: ($A_{ret} = A_p - A_r$)

Table A2.1 in Appendix 2 gives the theoretical forces of hydraulic cylinders in the English system of units. These figures do not account for the seal or packing friction in these cylinders. This type of friction is estimated to affect the thrust of the cylinders by about 10%.

Limitations on Maximum Thrust Force

The maximum thrust force which a cylinder can practically provide is limited by its piston-rod diameter and overall length. In cylinders with longer piston-rods, the piston-rod must be able to handle the thrust forces generated by the application. The cylinder must also be supported adequately.

Note that a head-end mounting provides greater column strength than the cap-end mounting, due to the smaller distance between the mounting points in the head-end mounting than that in the cap-end-mounting.

The piston-rod size of a hydraulic cylinder can be selected from the size charts, with the help of values of its free buckling length and the load imposed on the cylinder.

Example 8.1: A high-pressure double-acting hydraulic press cylinder with an effective piston area of 11 in² for push stroke, and a piston-rod area of 3.41 in², operating at 10153 psi does produce what theoretical forces for the push stroke and pull stroke? [1 ft² = 144 in²]

Solution

Effective piston area, push stroke, A_{push} = 11 in²
Piston Rod area, A_{rod} = 3.41 in²
Pressure, P   =10153 psi

Effective piston area, pull stroke, A_{pull} = A_{push} - A_{rod}
                                = (11 − 3.41) in² = 7.59 in²

Thrust, F_{push} = P x A_{push}
                     = (10153) x (11) = 111683 lb

Pull, F_{pull} = P x A_{pull}
                     = (10153) x (7.59) = 77061 lb

Cylinder Input Power: The hydraulic input power (P_{input}) supplied to the cylinder, in the English system units is given below:

$$P_{input} \text{ (hp)} = P \text{ (psi)} \times Q_A \text{ (gpm)} / 1714$$

Cylinder Output Power: The mechanical output power (P_{output}) of the hydraulic cylinder in the English system of units is given below:

$$P_{output} \text{ (hp)} = Force \text{ (lb)} \times Velocity \text{ (ft/s)} / 550$$

The input power supplied must be higher than the required output power to make allowances for losses on account of friction and leakage.

Cylinder Speed: Assume that the piston-rod assembly of a cylinder moves with a velocity of 'v' when pushed by the system fluid with a flow rate 'Q'. Further, assume that the cylinder piston of area 'A' has moved a distance 'S' in time 't' for attaining the velocity v. Figure 8.2 (a) and (b) shows two working positions of a cylinder with the piston in position 1 and position 2 respectively for determining the cylinder speed during its forward stroke. Figure 8.2(b) also shows the positions '1' superimposed. Mathematically,

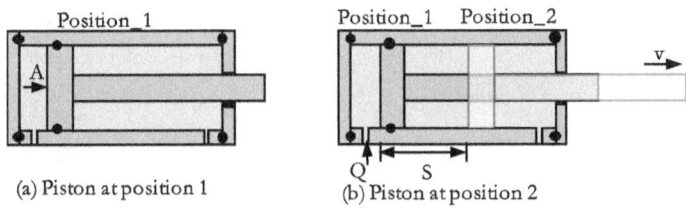

(a) Piston at position 1

(b) Piston at position 2

Figure 8.2 | Illustration of a cylinder in two piston positions

$$v = S/t \quad \text{or} \quad t = S/v$$

$$Q \text{ (ft}^3\text{/s)} = A \text{ (ft}^2\text{)} \times v \text{ (ft/s)}$$

We can easily relate the theoretical flow rate (Q) of the system fluid to the speed (v) at which the piston-rod moves if we consider the cylinder volume (V) that must be filled with the fluid and the distance (S) through which the cylinder piston must travel at the specified speed. The volume (V) of the cylinder is the length of the stroke (S) multiplied by the piston area (A).

The following section gives the flow rate (Q) to achieve the required speed (v), in the English units.

It can be observed from the equation mentioned above that the speed (v) of a given cylinder depends on the flow rate (Q) of the system fluid.

That is, a small-diameter cylinder moves faster as compared to a large-diameter cylinder with the flow rate remaining the same.

Examination of this equation also shows that the double-acting cylinder tends to produce somewhat a higher speed when retracting than when extending, provided that the system flow rate remains the same. This speed difference is mainly due to the different active areas exposed to the system fluid.

Operating Temperature
Seal compounds are designed for normal operating temperature from -5°F to +175°F. Ideally, the operating temperature should not exceed 110°F to 120°F to ensure the long life of the system.

Summary of Relations for Hydraulic Cylinders

Figure 8.3 gives the summary of essential relations of hydraulic cylinders, in the SI system units for the easy understanding and the correlation of these relations by the reader.

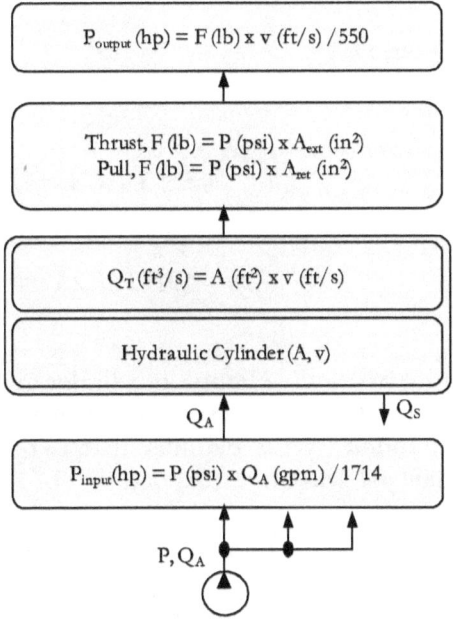

Figure 8.3 | Summary of relations - hydraulic cylinders

Appendix 3 presents typical cylinder data

Example 8.2
Determine the thrust of a hydraulic cylinder with a bore diameter 3.937 in, used for a flyover levelling application, and operating at a pressure of 10152 psi?

Solution

Bore diameter, D	= 3.937 in
Operating pressure, P	= 10152 psi

Piston area, extension stroke, $A = \pi D^2/4$
$$= \pi \times 3.94^2/4 = 12.17 \text{ in}^2$$

Thrust, F	$= P \times A$
	$= 10152 \times 12.17 = 123550 \text{ lb}$

Example 8.3
A double-acting hydraulic clamping cylinder must move out with a velocity 1.64 ft/s on the extension stroke. Calculate the flow rate demanded by the cylinder that is operating at a pressure of 3000 psi and producing a thrust of 11240 lb.

Solution

Thrust, F	= 11240 lb
Pressure, P	= 3000 psi
Speed, v	= 1.64 ft/s = 19.68 in/s

Area, A_{ext}	$= F/P$
	$= 11240/3000 = 3.75 \text{ in}^2$

Flow rate, Q	$= A_{ext} \times v$
	$= 3.75 \times 19.68 = 73.8 \text{ in}^3/s$

Chapter 9 | Hydraulic Motors

Some important terms relevant to the operation of hydraulic motors (Figure 9.1) are described below.

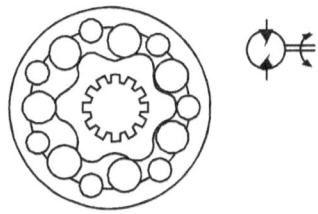

Figure 9.1 | A hydraulic motor

Operating Pressure (P): It is the pressure in a hydraulic system that overcomes all resistances in the system, which includes both useful work and losses. The rated pressure of a hydraulic motor is the maximum pressure, which the manufacturer recommends for the motor.

Motor Displacement (V_D): It refers to the volume of the system fluid required for turning the output shaft of a motor through one revolution. Some of the units of motor displacement are m^3/rev or cc/rev or in^3/rev.

Theoretical Flow Rate (Q_T): It is the quantity of the system fluid that must flow through a motor per unit of time, provided there is no leakage in the system. In the English system of units, the flow rate is measured in cubic-inch per minute (in^3/min) or gpm. The mathematical equations for the theoretical flow rate (Q_T) of the hydraulic motor in the English system of units are as follows:

$$Q_T(gpm) = \frac{V_D(in^3/rev) \times N(rpm)}{231}$$

Slippage in Hydraulic Motors: It is the internal leakage of the system fluid that passes through the unintended paths of a motor,

43

without performing any useful work. As the slippage in the hydraulic motor increases, more and more available flow intended for doing the useful work is lost, leading to the loss of power in the motor. However, all hydraulic motors are susceptible to some amount of slippage. The slippage increases, as the system pressure increases.

Speed: It is directly related to the theoretical flow rate to a motor and inversely related to the displacement of the motor. It can be expressed in the English system of units by the following equation:

$$\text{Speed, N (rpm)} = \frac{\text{Theoretical motor flow, } Q_T \text{ (gpm) x 231}}{\text{Motor displacement, } V_D \text{ (in}^3/\text{rev)}}$$

Therefore, it can be observed that, for a given flow rate, increasing the motor displacement decreases the motor speed and vice versa. Remember that the intended application decides the operating speed of the hydraulic motor.

Maximum motor speed is the speed of a hydraulic motor, at a particular inlet pressure, that it can sustain for a limited period without damage to the motor.

Minimum motor speed is the slowest, continuous, rotational speed obtainable from the output shaft of a hydraulic motor.

Input Power (P_{in}): Figure 9.2 gives the block diagram of a hydraulic motor, with the power relationships at its input side and output side.

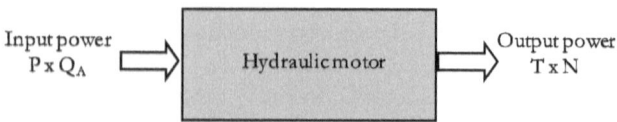

Figure 9.2 | A block diagram showing the power relationships in the hydraulic motor

The mathematical equation for the input power of the hydraulic motor, in the English system of units, is as follows:

$$\text{Input Horse Power (hp)} = \frac{P(\text{psi}) \times Q_{_A}(\text{gpm})}{1714}$$

Theoretical Torque (T_T): Theoretical torque of a hydraulic motor is a function of the motor's displacement and the system pressure. The following section gives the mathematical equation for the theoretical torque of the motor in the English system of units. The theoretical figures represent the torque available at the motor shaft, assuming no mechanical losses.

$$\text{Theoretical Torque, } T_{_T}(\text{in.lb}) = \frac{V_{_D}(\text{in}^3/\text{rev}) \times \Delta P(\text{psi})}{2\prod}$$

Breakaway (Starting) Torque of a hydraulic motor is the rotary force required for turning a stationary load connected to the motor. More torque is required to turn the stationary load than that required to keep it moving. This fact is because initially, the motor has to overcome the inertia of the load. Therefore, the motor needs a breakaway (starting) torque large enough to turn the load.

Running Torque of a hydraulic motor refers to the torque required to run a load connected to the motor. Remember, the running torque of the hydraulic motor changes whenever there is a variation in the associated system pressure.

Stalling Torque of a running hydraulic motor is the torque needed to stop the motor to a standstill.

Torque ripple of a hydraulic motor is the difference between the minimum torque and maximum torque delivered by the motor at a given pressure during its one cycle of rotation.

Example 9.1: A skid steer broom used to clean construction site has the hydraulic motor with a displacement of 3.2 in³/rev and operating at a pressure of 3000 psi. What is the maximum theoretical torque the motor is capable of producing?

Solution

Volumetric displacement, V_D = 3.2 in³/rev
Pressure, P   = 3000 psi

Theoretical Torque, T_T = V_D (in³/rev) x P (psi)/(2$\prod$) in.lb
$\quad\quad\quad\quad\quad\quad\quad$ = 3.2 x 3000 / (2π)
$\quad\quad\quad\quad\quad\quad\quad$ = 1527.89 in.lb

Actual Torque (T_A): It is the torque which a motor develops to drive the attached load alone. It is equal to theoretical torque minus the torque losses on account of any friction in the motor.

Output Power (P_{out}): The mathematical equation for the output power of a hydraulic motor, in the English system of units, is as follows:

$$\text{Output Horse Power (hp)} = \frac{T_A(\text{in.lb}) \times N \,(\text{rpm})}{63025}$$

Motor Efficiency: The efficiency of a hydraulic motor is the ratio of its output power and its input power. An ideal hydraulic motor is a motor that has no leakage and frictional losses, and it is 100% efficient. In practice, however, there are leakages and frictional losses taking place in the motor. Accordingly, two basic types of efficiencies are identified for the motor. They are: (1) Volumetric efficiency, and (2) Mechanical efficiency. Overall efficiency can, then, be derived from these two types of efficiencies. The following sections briefly explain these terms.

Volumetric Efficiency (η_v) of the hydraulic motor is the ratio of the theoretical flow rate responsible for developing the actual motor speed to the total flow rate consumed by the motor, including the leakage in the motor. Remember, the motor consumes more flow than it should theoretically, due to the leakage in the motor. The mathematical equation for the volumetric efficiency of the motor is as follows:

$$\text{Volumetric efficiency, } \left(\eta_v\right) = \frac{\text{Theoretical flow rate } (Q_T)}{\text{Actual flow rate } (Q_A)}$$

Remember, the speed of a hydraulic motor is entirely dependent on the flow through the motor, and is independent of the pressure drop across the motor.

Mechanical efficiency (η_m) of the hydraulic motor is the ratio of the actual torque delivered by the motor to the theoretical torque of the motor. The hydraulic motor produces less torque than it should theoretically, due to the frictional losses in the motor. The mathematical equation for the mechanical efficiency of the hydraulic motor is as follows:

$$\text{Mechanical efficiency, } \left(\eta_m\right) = \frac{\text{Actual torque, } (T_A)}{\text{Theoretical torque } (T_T)}$$

Overall Efficiency (η_o) of the hydraulic motor is the ratio of the 'brake' power delivered by the motor to the hydraulic power delivered to the motor. Both volumetric and mechanical efficiencies reduce the overall performance of the motor. Therefore, the overall efficiency of the motor is also the product of its volumetric efficiency and its mechanical efficiency and is expressed mathematically as:

$$\text{Overall efficiency, } \left(\eta_o\right) = \frac{\text{Brake power delivered by the motor}}{\text{Hydraulic power delivered to the motor}}$$

$$= \eta_v \text{ x } \eta_m$$

Summary of Relations for Hydraulic Motors

Figure 9.3 gives the summary of essential relations of hydraulic motors, in the SI system units, for the easy understanding and the correlation of these relations by the reader.

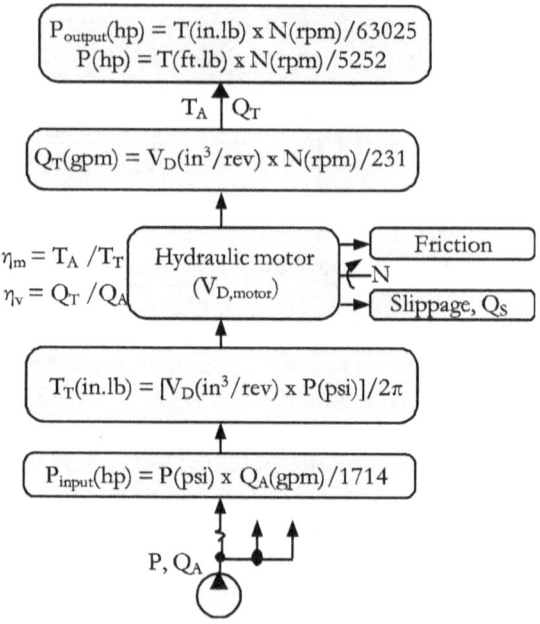

Figure 9.3 | Summary of relations for a hydraulic motor

Appendix 2 presents typical hydraulic motor data.

Example 9.2

A hydraulic motor rotates at a speed of 450 rpm with a nominal displacement of 0.244 in³/rev. The pressure differential across the hydraulic motor is 1088 psi. The overall efficiency is 80%, and the volumetric efficiency is 90%. Calculate the following: (1) Theoretical flow rate, (2) Actual flow rate, (3) Power input, (4) Shaft power and (5) Shaft torque.

Solution

Speed, N	= 450 rpm
Displacement, V_D	= 0.244 in³/rev
ΔP	= 1088 psi
η_v	= 90%
η_o	= 80%
Speed, n	= 450/60 rps = 7.5 rps
Theoretical flow rate, Q_T	= V_D (in³/rev) x N(rpm)/231
	= (0.244 x 450)/231
	= 0.475 gpm
Actual flow rate, Q_A	= Q_T/ η_v
	= 0.475/0.9 = 0.528 gpm
Power input P_{in}	= ΔP (psi) x Q_A (gpm)/1714
	= 1088 x 0.478/1714
	= 0.3 hp
Shaft power, P_{out}	= P_{in} x η_o
	= 0.3 x 0.8 = 0.243 hp
Shaft torque, T	= P_{out} (watt) /2π n
	= (P_{out} (hp) x63025)/N(rpm)
	= (0.243x63025)/450
	= 34 in.lb

Chapter 10 | Flow Rate Coefficient of Control Valves

When fluid flows through a hydraulic valve (Figure 10.1), the pressure across the valve reduces. A hydraulic valve needs to be rated according to the ability of the valve to pass the fluid through it, keeping the pressure drop and the energy loss within limits. Every valve is assigned a unique capacity index by its manufacturer to measure the pressure drop across the valve and to make a fair comparison amongst similar valves of different types and makes. This index is also known as the flow-rate coefficient or the valve-sizing coefficient.

Figure 10.1 | A hydraulic valve

The flow coefficient of a valve describes the relationship between the pressure drop (ΔP) across the valve with the flow rate (Q) through the valve. The flow coefficient is referenced for water at the specific operating conditions. At the same time, the water formulas apply to ordinary liquids also. The flow coefficient is measured using the standard test setup, as shown in Figure 10.2.

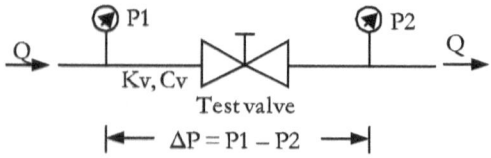

Figure 10.2 | A test setup to calculate flow coefficient

Flow coefficient of the valve can be expressed in several ways. They are:

- Cv is the flow coefficient term used for measuring the capacity of the valve in English units. It is defined as the flow rate (gpm) of the water flowing through the valve at the temperature of 60°F that causes one psi pressure drop across it.

- Kv is the flow coefficient term used for measuring the capacity of the valve in SI system units. It is defined as the flow rate (m^3/h) of water flowing through the valve at a temperature in the range from 5 to 30°C that causes one bar pressure drop across it.

The general definition of the flow coefficient can be expressed as equations for modelling the flow of liquids. These equations are given below:

$$Q = Cv \times \sqrt{(\Delta P / SG)}$$

Where,
Q is the flow rate in gpm
ΔP is the pressure drop across the valve in psi
Cv is the flow coefficient in 'gpm/$\sqrt{psi}$'
SG denotes the specific gravity

The value of flow coefficient Cv for a flow control valve in its fully open position is determined experimentally and is listed as the rated C_v in its manufacturer's catalogues.

Chapter 11 | Hydraulic Accumulator Sizing

Gas-charged accumulators are widely used in many industries. They are typically rated in terms of the gas volume they can have when all the fluid has been discharged. Accurately sizing an accumulator for a hydraulic system is a difficult task, even when only one cylinder is involved in the system. To calculate the capacity of the accumulator for the system requires intimate knowledge of the operating conditions of the system. Then, using any one of the mathematical models, as explained below, the size of the accumulator can be calculated. Manufacturers also bring out sophisticated software packages for the accurate determination of the capacity of an accumulator.

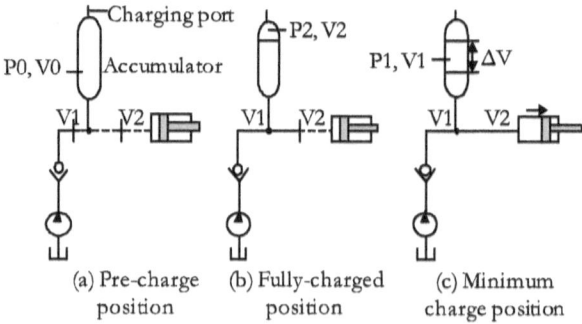

(a) Pre-charge position   (b) Fully-charged position   (c) Minimum charge position

Figure 11.1 | Three positions of a gas-charged accumulator connected in a hydraulic system

Figure 11.1 shows different positions of a hydraulic system with a pump, a cylinder, a check valve, and a piston-type gas-charged accumulator for supplementing the power. Let 'D' be the bore diameter and 'S' be the stroke length of the cylinder. Figure 11.1 (a) shows the position of the accumulator when it is almost pre-charged. Let P_0 [= (0.95 to 0.97) x P_1 (i.e., Minimum system pressure)] be the pre-charge pressure in the accumulator and V_0 be its corresponding volume. V_0 is the required accumulator size.

Figure 11.1(b) shows the position of the hydraulic circuit when the pump is turned on keeping the DC valve V_1 open, and the valve V_2 closed. During this period, the accumulator gets charged to the maximum pressure corresponding to the setting of the pressure relief valve (not shown) in the system. Let P_2 be the maximum charge pressure in the accumulator and V_2 be the corresponding volume during this stage. Figure 11.1(c) shows the position of the circuit when both the DC valves V_1 and V_2 are open, and the cylinder has just reached the end of its stroke. During this period, the accumulator discharges fluid into the system, and the pressure in the accumulator decreases. Let P_1 be the pressure in the accumulator during this stage and V_1 be the corresponding volume. It may be noted that the fluid capacity available in the accumulator to drive the cylinder is $(V_1 - V_2)$.

A gas-filled accumulator can be charged or discharged under the isothermal or the adiabatic conditions. In the isothermal process, the compression and expansion of the gas in the accumulator take place slowly, so that the existence of an approximately constant temperature condition can be presumed, as the complete heat transfer is possible between the gas medium and the environment. In the adiabatic process, the compression and expansion of the gas are quick so that no heat transfer is possible between the gas medium and the environment. The required accumulator size for various charging and discharging conditions can be derived in the following ways.

Accumulator Charging and Discharging under Isothermal Condition: The perfect gas laws govern the compression and decompression of the nitrogen gas contained in the accumulator. Using the Boyle-Mariotte's law for ideal gases, assuming slow charging or slow discharging, to allow the gas in the accumulator to maintain its temperature close to a constant, we get the size of accumulator V_0 from the following equations:

$$P_0 V_0 \;=\; P_1 V_1 = \; P_2 V_2$$
$$\text{That is, } V_1 - V_2 = [(P_0 V_0) / P_1] - [(P_0 V_0) / P_2]$$
$$= V_0 [(P_0 / P_1) - (P_0 / P_2)]$$

Therefore,
Accumulator volume, $V_0 = [V_1 - V_2] / [(P_0/P_1) - (P_0/P_2)]$

$(V_1 - V_2)$ is the volume of the fluid required for the full extension of the cylinder.

$$\text{That is, } V_1 - V_2 = (\prod D^2/4) \times S$$

As the gas is pressurized, its temperature is liable to go up, and the volume of the fluid entering the accumulator is lower than the calculated amount. This deficiency can be compensated by increasing the accumulator capacity by about 5%. If there are many actuators in the system, consider the peak loading when sizing the accumulator.

Accumulator Charging and Discharging under Adiabatic Condition: Most fluid power designers use the ideal gas laws for the design calculations of accumulators. However, the primary gas laws do not apply when there is no or little heat transfer into or out of an accumulator, as found in a shock absorber or a pulsation damper or an emergency power source. Remember, today's hydraulic systems move faster with higher cycle rates. There exists short time for heat to enter or leave the accumulator, so we assume that the compression and expansion of the gas be adiabatic – that is; no heat is transferred into or out of the accumulator. Assuming quick charging and quick discharging, we get the size of accumulator V_0 from the following complicated equations:

$$P_0 V_0^n = P_1 V_1^n = P_2 V_2^n$$

Where, n is the polytropic exponent, which is about 1.4 for a diatomic gas.
Similarly, the size of the accumulator with charging and discharging under the adiabatic condition as:

$$V_{0,\text{adiabatic}} = [V_1 - V_2] / [(P_0/P_1)^{1/n} - (P_0/P_2)^{1/n}]$$

Accumulator with Slow Charging and Quick Discharging: Accumulator capacity when the charging process is slow (isothermal), and the discharging process is quick (adiabatic) can be calculated from the following equation:

$$V_0 = [V_1 - V_2] / \{(P_0 / P_2)^{1/n} - [(P_2 / P_1)^{1/n} - 1]\}$$

Temperature Influence, Accumulator Sizing: Any variation in the temperature should be taken into consideration when the accumulator volume is calculated. Let the capacity of the accumulator be V_0 at the temperature T_1 (K), and the capacity of the accumulator has increased to V_{oT} when the temperature has risen to T_2 (K). The capacity (V_{oT}) can be calculated by using the following formula:

$$V_{oT} = V_0 \times (T_2/T_1)$$

Correction Coefficient at Higher Pressures: The nitrogen gas in accumulators does not behave according to the ideal gas laws when the pressure goes beyond 200 bar. The capacity of the accumulator (V_{oP}), at higher pressures, under isothermal and adiabatic conditions, can be calculated by using the following formula:

$$V_{oP} = V_0 / C_i \text{ (for isothermal condition)}$$
$$= V_0 / C_a \text{ (for adiabatic condition)}$$

Where C_i is the isothermal correction coefficient, and C_a is the adiabatic correction coefficient, both have to be deduced from the standard charts published by manufacturers.

Chapter 12 | Fluid Conductors

In a conventional hydraulic system, various components of the system are assembled through a conductor system. The conductor system is a network of pipes, tubing, and hoses that connects to the components through fittings for the effective delivery of the fluid through the system.

A conductor is a pressure-tight vessel used to convey a sufficient quantity of pressurized fluid through it in a leak-free manner. It must have smooth interiors to reduce the friction in sliding parts of system components and flow turbulence during the fluid flow, and sufficient wall thickness to withstand the high operating and shock pressures developed in the system. Next, it must be capable of withstanding the high system as well as ambient temperatures. Further, it must also be compatible with the type of fluid used.

The critical considerations for the selection of fluid conductors include their construction, sizing, installation, routing, and applicable standards.

Terms and Definitions, Fluid Conductor
The fluid power industry uses a multitude of conductor-related terms to specify the performance levels of fluid power systems. For example, a fluid conductor is specified by its diametrical size and wall thickness. The following sections present a brief account of some commonly used terms and definitions of fluid conductors.

Diametrical Size, Fluid Conductor
The diametrical size of a conductor is specified by its inside diameter, outside diameter, or nominal size (Figure 12.1).

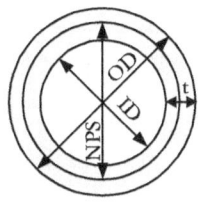

Figure 12.1 | Size specifications of a fluid conductor

Inside Diameter, D_i, Fluid Conductor
It is the smallest cross-sectional diameter of a conductor.

Outside Diameter, D_o, Fluid Conductor
It is the largest cross-sectional diameter of a conductor.

Nominal Size, Fluid Conductor
In ANSI and SAE standards, for example, the size of a pipe is specified in terms of nominal pipe size (NPS), and in SI system, it is specified in terms of Nominal Diameter (DN).

Wall Thickness, t, Fluid Conductor
The wall thickness of a pipe or tubing decides the maximum pressure that can be subjected to it. It is expressed in terms of a schedule number in ANSI and SAE standards or metric units. It is given by:

$$\text{Wall thickness, } t = (D_o - D_i) / 2$$

Schedule Number
In ANSI/SAE system, the wall thickness of a pipe is usually described in terms of a 'schedule number'. The schedule numbers vary from 5 through 160 in a graded manner. Altogether, there are eleven different schedule numbers. They are: 5, 10, 20, 30, 40, 60, 80, 100, 120, 140, and 160.

Larger schedule number for a given pipe size points to more substantial wall thickness. For a particular size of the pipe, the outer diameter stays the same, but the inside diameter becomes smaller as its schedule number increases.

As an example, Table 1.1 gives the wall thicknesses corresponding to schedule numbers 40, 80 and 160, for a pipe of nominal size ½.

Table 1.1 | Wall thicknesses

Nominal Pipe Size		Outside Diameter	Wall thickness		
			Schedule 40	Schedule 80	Schedule 160
		inch	inch	inch	inch
½	0.500	0.840	0.109	0.147	0.188

Example 12.1
Determine the inside diameter of a pipe with an OD of 1.9 inch and a wall thickness of 0.145 inch.

Solution

Pipe OD	= 1.9 inch
Wall thickness	= 0.145 inch
Inside diameter of the pipe	= 1.9 – (2 x 0.145) inch
	= 1.61 inch

Hoop Stress, Fluid Conductor
Hoop stress in a pipe or tubing is the circumferential stress acting on the wall of the conductor that is trying to split it. It is the maximum pressure that the material of the conductor is capable of withstanding before pulling apart.

Consider a thin-walled conductor of the inside diameter (D_i), wall thickness, (t, where $t < 0.1 \times D_i$), and length (L), as shown in Figure 12.2. The conductor is subjected to operating pressure, P.

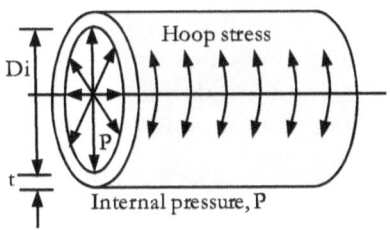

Figure 12.2 | Hoop stress

The operating pressure, acting normal to the inside surface of the pipe, induces a circumferential force in the conductor that tends to split the conductor into two halves. The normal surface area can be taken as the projected area (D_i x L) of one half of the pipe. The circumferential force (or burst force) is given by:

$$\text{Circumferential force} = P \times D_i \times L$$

The tensile force which is trying to resist the splitting of the conductor acts on the cross-sectional area (tL) of each wall. Therefore, the resistive force is given by:

$$\text{Resistive force} = 2tL \times \text{Hoop stress}.$$

Equating the circumferential force to the resistive force, we get,

$$P \times D_i \times L = 2tL \times \text{Hoop stress}$$

Therefore,

$$\text{Hoop stress} = P \times D_i / 2t$$

The conductor must have sufficient tensile strength to prevent its bursting due to the excessive hoop stress.

Example 12.2
Determine the hoop stress developed in a pipe of outside diameter 2.375 inch and a wall thickness of 0.218 inch when the pipe is subjected to a pressure of 1000 psi.

Solution

Outside diameter of the pipe, Di	= 2.375 inch
Wall thickness of the pipe, t	= 0.218 inch
Pressure, P	= 1000 psi
Hoop stress developed	= P x D_i / 2t
	= 1000 x 2.375 / (2 x 0.218)
	= 5447 psi

Burst Pressure, Fluid Conductor
It is the internal pressure inside a fluid conductor that causes it to burst or rupture (Figure 12.3). The fluid conductor bursts when the hoop stress exerted on the conductor exceeds the tensile strength (S) of the conductor material. Barlow's formula is commonly used to predict the burst pressures in ductile thin wall tubes (t < 0.1 x D_i)

$$\text{Burst pressure (BP)} = 2tS / D_i$$

For thick-walled pipes, the tensile stress across the wall thickness is not uniform. Therefore, the following formula must be used to take into account the non-uniform tensile stress.

$$\text{Burst pressure (BP)} = 2tS / (D_i + 1.2t)$$

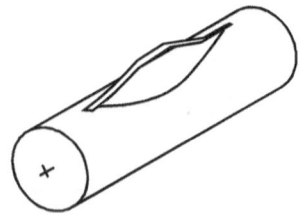

Figure 12.3 | Burst tested tube

Working Pressure, Fluid Conductor

The working pressure of a conductor is the safe pressure to which it can be subjected. It is calculated by dividing the burst pressure of the conductor with a safety factor.

$$\text{Working pressure (WP)} = \frac{\text{Burst pressure (BP)}}{\text{Safety factor (SF)}}$$

- A safety factor of 4:1 is used for hydraulic applications where shock and mechanical strain are not considerable.
- A safety factor of 6:1 should be used where considerable shock and mechanical strain are expected.
- A safety factor of 8:1 should be used where severe hydraulic shock and mechanical strain are expected.

Design Pressure, Fluid Conductor

It is the pressure to which each component of a piping system is designed. It is not to be less than the actual pressure at the most severe condition of pressure and temperature expected during the service of the piping system.

Maximum Allowable Working Pressure, Fluid Conductor

It is the maximum pressure of a piping system, determined by the weakest component of a piping system. It is not to exceed its design pressure.

Minimum Bend Radius, Fluid Conductor

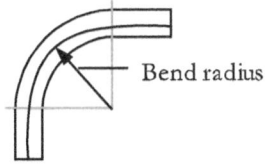

Figure 12.4 | Minimum bend radius

61

It is the smallest radius of the curved section of a conductor (tube or hose) beyond which it should not be bent without flattening, kinking or wrinkling (Figure 12.4). Bending of the conductor beyond the limit causes severe backpressure and damages the conductor internally leading to its premature failure.

Design Temperature
It is the maximum temperature at which each piping component is designed to operate. It may be taken as the maximum fluid temperature.

Example 12.3
Determine the maximum pressure rating of a thick-wall pipe of OD of 1.315 inch with a wall thickness of 0.25 inch. The pipe material is carbon steel with a tensile strength of 50000 psi. Assume a safety factor of 4.

Solution:

Pipe OD, D_o	= 1.315 inch
Wall thickness, t	= 0.25 inch
Tensile strength, S	= 50000 psi
Safety factor, SF	= 4

Pipe ID, D_i
$$= D_o - 2 \times t$$
$$= 1.315 - (2 \times 0.25)$$
$$= 0.815 \text{ inch}$$

Burst pressure
$$= 2ts / (D_i + 1.2t)$$
$$= 2 \times 0.25 \times 50000 / [0.815 + (1.2 \times 0.25)]$$
$$= 22422 \text{ psi}$$

Pressure rating
$$= \text{Burst pressure/Safety factor}$$
$$= 22422 / 4 = 5605 \text{ psi}$$

Fluid Velocity

It is the average velocity (v) at which fluid particles moves past a given cross-section (A) in a conductor. The relation between the flow rate (Q) and the fluid velocity is given by:

$$Q \ (ft^3/s) = A(ft^2) \times v(ft/s)$$

The relation for the fluid velocity in commonly used units are given by:

$$v(ft/s) = \frac{0.3208 \times Q \ (gpm)}{A \ (in^2)}$$

Example 12.4
The fluid is flowing through a pipe of an internal diameter of 1" size at the rate of 20 gpm. Determine the fluid velocity.

Solution

Pipe size	= 1"
Flow rate	= 20 gpm

Area of cross-section $= \pi \cdot D^2/4$
$= 3.14 \times 1 \times 1/4 = 0.785 \ in^2$

$$v(ft/s) = \frac{0.3208 \times Q \ (gpm)}{A \ (in^2)}$$

$$v(ft/s) = \frac{0.3208 \times 20}{0.785}$$

$$= 8.17 \ ft/s$$

Pipe and Tube Materials

The pipe and tube materials suitable for high-pressure industrial hydraulic service are: (1) Cold-drawn seamless carbon steel and (2) Cold-drawn seamless stainless steel. Stainless steel pipes and tubes are used in applications that require resistance to corrosion, such as in chemical equipment or marine vessels. Hot rolled pipes are not recommended for hydraulic services as they leave behind scales inside and outside.

Steel is an alloy of iron and carbon ($<2\%$). Steels are classified into four basic types according to the amount of carbon and other alloys. They are carbon steel, alloy steel, stainless steel, and tool steel.

- Carbon steel is an iron-carbon alloy. It also contains manganese, silicon, and copper, lead, aluminum, cobalt, chrome, nickel, etc. in varying degrees. Carbon steels are vulnerable to corrosion.

- Alloy steel is prepared by adding steel with various metals such as iron, nickel, aluminum, copper, etc. The strength and properties of alloy steel depend on the amount of the elements present in the alloy steel.

- Stainless steel has low carbon content but contains chromium alloy, and nickel or molybdenum. It is strong and corrosion-resistant, and it can withstand high temperatures.

- Tool steels are hard and used to make metal tools and mold-making tools.

Steel specifications are given by different organizations like ISO, AISI, SAE, ASTM, European Standard (EN), German standard (DIN), Japanese Industrial Standards (JIS), etc. in their unique ways.

Properties and standards of typical steel materials

Table 1.2 | Properties and standards of typical steel materials

Pipe material	Properties	Standards
Cold-drawn seamless carbon steel	High-pressure capability, precise dimensions/shape, clean inside surface with no scale, excellent scaling surface after roll flaring	DIN EN 10305-4 E 355N (St. 52.4 NBK) E 235N (St. 37.4 NBK)
Cold-drawn seamless stainless steel	High-pressure capability, precise dimensions/shape, excellent scaling surface after roll flaring	DIN EN 10216-5 ASTM A269/A213 ASTM A312

Tensile Strengths and Yield Strengths of Steels

Table 1.3 | Tensile strengths and yield strengths of steels

Steel type	Tensile strength (N/mm^2)	Yield strength $(N/mm^2$ min)
E235N tubes (St 37.4)	340	235
E355N tubes (St 52.4)	490	355
AISI 316L metric size tubes	485	170
TP 316L schedule size pipes	485	170
DIN 2391 St. 45	570	255
DIN 2391 St. 52	630	355

E – Steel for machine parts
235 – Minimum yield strength in N/mm^2

Appendix 4 presents typical fluid conductor data.

Chapter 13 | Design of Hydraulic Piping Systems

A piping system for an application is to be designed in conformity with many requirements of the application. In general, the suction pipe should be short and straight, and the return line should be large enough to limit the backpressure. When designing the piping system, the following points have to be taken into account: (1) System pressure, (2) Operating temperature, (3) Duty cycle, (4) Shocks and vibration, (5) Material, (6) Connection technology, (7) Hoses and hose couplings, (8) Pipe supports, and (9) Standards.

Fluid Velocities – Suction Lines
The suction line is typically dimensioned so that the velocity does not exceed 1.2 m/s. The recommended fluid velocities to be utilized for initial pipe sizing in suction lines are given in Table 13.1.

Table 13.1 | Fluid velocities in suction lines

Viscosity (cSt)	Maximum velocity (ft/s)
150	2
100	2.5
50	3.6
30	4

Fluid Velocities - Pressure Lines
The recommended fluid velocities for initial pipe sizing in pressure lines are given in Table 13.2.

Table 13.2 | Fluid velocities in pressure lines

Pressure line	For flow rate >2.6 gpm
900 – 1450 psi	13 – 14
1450 – 2300 psi	14 – 16
2300 – 3600 psi	16 – 18
3600 – 5800 psi	18 – 20

Fluid Velocities – Return Lines

Fluid velocities to be utilized for initial pipe sizing in return lines should be between 6.5 to 10 ft/s.

Dimensioning Based on Flow Velocity

When using the dimensioning method based on flow velocity, the inner diameter of the pipe can be determined by using the equation below, when a maximum flow rate and recommended flow velocity are known.

$$d= \sqrt{\frac{4 \times Qmax}{\Pi \times v}}$$

d   =Inner diameter of the pipe (ft)
Q_{max} =Maximum flow rate (ft³/s)
v   =Flow velocity (ft/s)

Example 13.1

Determine the inner pipe diameter to create a flow velocity of 15 ft/s with a flow rate of 0.0589 ft3/s.

Solution

Maximum Flow rate, Qmax   = 0.0589 ft³/s
Velocity, v   = 15 ft/s

$$Pipe\ ID = \sqrt{\frac{4 \times Qmax}{\Pi \times v}}$$

$$= \sqrt{\frac{4 \times (0.0589)}{\Pi \times 15}}$$

$$= 0.0707\ ft = 0.8484\ in$$

Dimensioning based on Pressure Losses

When using the dimensioning method based on the pressure losses, the inner diameter of the pipe is selected so that the resulting pressure losses do not increase above a specified value. The total pressure loss is allowed to be 3 to 5% for systems in continuous use. The total pressure loss is allowed to be 7 to 10% for systems with intermittent duty cycle.

Flow Types

The amount of pressure losses also depends on the type of flow. When the flow is laminar, all the fluid particles move parallel to the pipe. When flow velocity increases, the flow will become turbulent, which means the direction of the individual fluid particles varies. The flow type can be found by determining the so-called Reynolds number (Re) and comparing it to critical Reynolds number value. The Reynolds number can be determined with the equation:

$$Re = \frac{v \cdot d \cdot \varrho}{\mu} = \frac{v \cdot d}{v}$$

Where,

Re = Reynolds number [-]
v   = Flow velocity [ft/s]
d   = Inner diameter of the pipe [in]
v (nu) = Kinematic viscosity [ft^2/s]
μ = Absolute viscosity [lb.s/ft^2]
ϱ = Density of the fluid [slug/ft^3]

The flow is said to be laminar when Re < Re (critical) and the flow may be treated as turbulent when Re > Re (critical). Remember, Re (critical) = 2000

The Reynolds number, with the kinematic viscosity in cSt, is given by:

$$Re = \frac{7740 \times v(ft/s) \times d(in)}{v \ (cSt)}$$

68

Example 13.2

Determine the inner pipe diameter to create a flow velocity of 15 ft/s with a flow rate of 0.05886 ft³/s. A fluid with a density of 1.6 slug/ft³ and an absolute viscosity of 0.00668 lb.s/ft² is flowing through the pipe. Is the flow laminar or turbulent?

Solution

Q	$= 0.05886$ ft³/s
v	$= 15$ ft/s
μ	$= 0.00668$ lb.s/ft²
ϱ	$= 1.6$ slug/ft³
A	$= Q/v = 0.05886 / 15 = 0.003924$ ft²
d	$= \sqrt{(4 \times A /\pi)} = 0.0707$ ft $= 0.8484$ in

$$\text{Reynolds number, Re} = \frac{v \cdot d \cdot \varrho}{\mu}$$

$$Re = \frac{15 \times 0.0707 \times 1.6}{0.00668}$$

$$Re = 254 \ (<2000, \text{ Laminar flow})$$

Example 13.3

If fluid with a kinematic viscosity of 30 cSt is flowing at a velocity of 10 ft/s through a pipe of ¾-inch diameter, what would be the Reynolds number?

Solution

ν	$= 30$ cSt
v	$= 10$ ft/s
d	$= 0.75$ in

$$Re = \frac{7740 \times v(ft/s) \times d(in)}{\nu \ (cSt)}$$

$$Re = \frac{7740 \times 10 \times 0.75}{30}$$

$$Re = 1935 \ (\text{Laminar flow})$$

Dimensioning Based on Pressure Losses
The overall pressure losses in the piping include frictional pressure losses arising in straight pipe sections, as well as individual pressure losses arising in bends and junctions.

Frictional Pressure Losses
The pressure losses in pipes and hoses can be estimated from the Darcy-Weisbach equation as given below:

$$\Delta Pa = \lambda \cdot \frac{1}{d} \cdot \frac{\varrho\, v^2}{2}$$

Δp_a	= Frictional pressure loss, [lb/ft²]
λ	= Frictional resistance factor [-]
l	= Length of the pipe [ft]
d	= pipe id [ft]
ϱ	= Hydraulic fluid density [Slug]
v	= Flow velocity [ft/s]

$$[1 \text{ lb/ft}^2 = 1/144 = 0.00694444 \text{ psi}]$$

Frictional Resistance Factor, λ
The friction factor is a function of Reynolds number and the surface roughness of round pipes and can be determined mathematically or by consulting the classic Moody Diagram. For laminar flow, the friction factor is independent of the surface roughness, and for turbulent flow, the friction factor is dependent both on the Reynolds number and the surface roughness.

For Laminar Flow
If the flow is laminar, λ depends only on the Reynolds number. The friction factor is given by:

$$\lambda = \frac{64}{Re}$$

Frictional Pressure Losses in Straight Pipe Sections

The pressure loss in a pipe for laminar flows can be determined by using the following equations in terms of the average velocity of flow and the more favorable fluid flow rate, respectively:

$$\Delta Pa = \frac{32 \, \mu \, l \, v}{d^2}$$

$$\Delta Pa = \frac{128 \, \mu \, l \, Q}{\pi \, d^4}$$

Note: The pressure drop scales linearly with line length, therefore, long lines, and smaller diameter lines, could impact system efficiency.

Example 13.4

Calculate the frictional pressure loss for a 2.067-inch internal diameter pipe of length 100 ft through which fluid is flowing at a velocity of 3.8 ft/s. Kinematic viscosity = 44.4 cSt, and the density of the fluid 1.6 slug/ft³.

Solution

v	= 3.8 ft/s
d	= 2.067 inch = 0.17225 ft
ν	= 44.4 cSt
ϱ	= 1.6 slug/ft³
l	= 100 ft
Re	= 7740 . v . d / ν =7740 x 3.8 x 2.067 /44.4
	= 1369
λ	= 64/1369 = 0.0467
Δp_a	= λ (l/d) (ϱ.v²/2)
	= 0.0467 x (100/0.17225) x (1.6 x 3.8²/2)
	=0.0467 x 580.55 x 11.552
	=313 lb/ft² = 313/144 psi = 2.17 psi

Frictional Losses in Turbulent Flow

The inside surface of a round pipe is given in Figure 13.1. The mean height of the roughness is designated as 'ε' and 'd' is the pipe inside diameter.

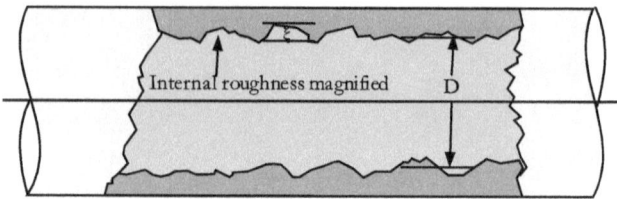

Figure 13.1 Relative roughness in a pipe

Relative roughness of the pipe's inside surface is defined as the mean roughness (ε) divided by the pipe inside diameter (d). That is,

Relative roughness = ε/d

Pipe roughness values depend on the pipe material as well as the method of its manufacture.

Typical values of absolute roughness:

Drawn tubing   – 0.000059 in
Cast iron   – 0.01 in
Riveted steel   – 0.07 in

Moody Diagram

The Moody Diagram, as given in Figure 13.2, plots the friction factor as a function of the Reynolds number and the relative pipe roughness. From the diagram, it can be inferred that for laminar flows, the friction factor depends only on the Reynolds number. For turbulent flows, the friction factor depends both on the Reynolds number and the relative roughness.

The following procedure can be followed to find the friction factor from the Moody Diagram:

- Find the value of Re
- Find the relative roughness ε
- Project a vertical line on the Re axis at the value of Re determined
- Find the curve corresponding to relative roughness
- Project horizontally to the f axis to obtain the friction factor

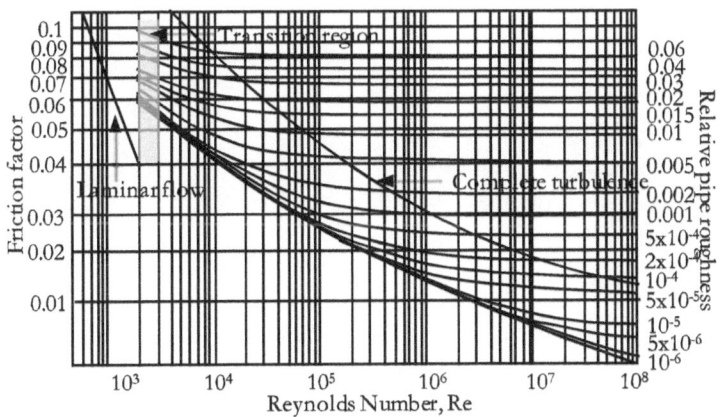

Figure 13.2 | Moody Diagram
Ref: Moody, L F (1944), 'Friction factors for pipe flow'

Individual Pressure Losses

The individual pressure losses occur in pipe bends, junctions and generally in pipe sections where the cross-sectional area or flow direction changes. The relation for the individual pressure losses is given by:

$$\Delta Pb = \zeta \frac{\varrho \cdot v^2}{2} = \zeta \frac{\varrho \cdot Q^2}{2 A^2}$$

Where,

Δp_b = individual pressure loss [lb/ft^2]

ζ = individual resistance factor [-]

ϱ = the fluid density [slug]

v = flow velocity [ft/s]

Values of Loss Coefficient (ζ)

The value of the individual resistance factor ζ depends on the flow channel structure and dimensioning:

The loss coefficients (ζ):

- 90^0 elbow $\quad\quad\quad$ -0.2
- 45^0 elbow $\quad\quad\quad$ -0.15
- Tee fitting $\quad\quad\quad$ -0.9
- Sharp-edged entrance -0.5
- Rounded entrance $\quad$ -0.05
- Sharp-edged exit $\quad$ -1.0
- Rounded exit $\quad\quad$ -1.0

Individual pressure losses can also be found from the nomogram given in Figure 13.3

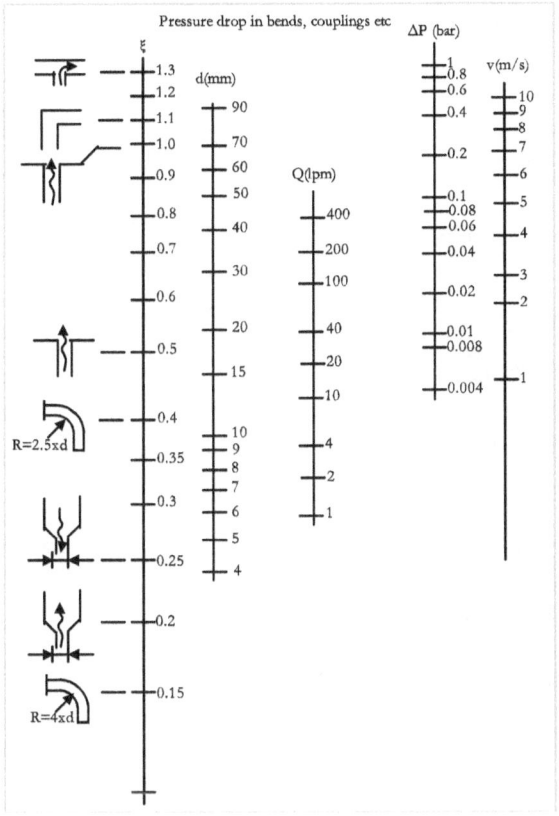

Figure 13.3 | Individual pressure losses

Total Pressure Losses

The total pressure losses in the piping are obtained by adding up the frictional pressure losses and the individual pressure losses.

$$\Delta Ptot = \Delta Pa + \Delta Pb$$

The total pressure loss is allowed to be 3 to 5% for systems in continuous use. The total pressure loss is allowed to be 7 to 10% for systems with intermittent duty cycle.

Chapter 14 | Steps for Hydraulic System Design

A properly designed hydraulic system is an interconnection of correctly-sized components, such as a power pack, pressure relief valve, actuators, control valves, filters, and heat-exchangers using pipes, tubes and hoses.

The design of a hydraulic system involves the following basic steps:
- selection and sizing of components,
- determining the system operating pressure and flow rate, and
- finding the component specifications to meet the design objectives.

There are many possible solutions for designing a hydraulic system as there are many components available in the market with varying specifications and quality.

Optimum design of a hydraulic system for a project must try to synchronise the specifications and quality of components required with the specifications and quality of components available in the market.

Further, the designer must take into account the fund available for implementing the project.

Here, a few examples of designing hydraulic systems are presented, purely for educational purpose.

You may apply the knowledge to design real systems by taking into account the site conditions. That is, the design must take place after a thorough system analysis.

Critical Design Steps

The following critical steps may be followed for finding the significant parameters while designing a hydraulic system with a pump and cylinder. Similar steps may be followed for designing a hydraulic system with a pump and a hydraulic motor.

An analysis of the system to be designed would reveal the application requirements of output force (F) or torque, speed (v), and output power (P_{out}).

For example, these parameters for a cylinder are governed by the relation:

$$P_{out}(kW) = F(N) \times v(m/s)/1000$$

The volumetric efficiency (η_{vc}) and mechanical efficiency (η_{mc}) of the cylinder can be assumed as per the intended quality of the cylinder, and its overall efficiency (η_{oc}) can be calculated from the relation

$$\eta_{oC} = \eta_{vC} \times \eta_{mC}$$

The volumetric efficiency (η_{vp}) and mechanical efficiency (η_{mp}) of the pump can be assumed as per the intended quality of the cylinder, and its overall efficiency (η_{op}) can be calculated from the relation

$$\eta_{oP} = \eta_{vP} \times \eta_{mP}$$

Next, the hydraulic power (P_{hyd}) involved in the hydraulic power transmission system can be calculated using the following equation:

$$P_{hyd} = P_{out} / \eta_{oC}$$

The mechanical power input to the pump (P_{input}) corresponds to the electric motor power rating, and the required input power can be calculated using the following equation:

$$P_{input} = P_{hyd} / \eta_{oP}$$

The flow rate (Q_A), the maximum pressure rating (P_{max}), and prime mover speed (N_p) of the pump can be selected from the manufacturer's datasheet.

The actual pump flow rate (Q_{AP}) is the same as the actual cylinder flow rate (Q_{AC}) and can be taken as (Q_A).

Calculate the theoretical flow rate of the cylinder (Q_{Tc}) from the following equation:

$$Q_{TC} = Q_A \times \eta_{vC}$$

Find the piston area (A_{ext}) of the cylinder from the following equation.

$$A_{ext} = Q_{TC} / v$$

Find the piston diameter (D) of the cylinder from the following equation and reconcile with the data from the manufacturer's domain:

$$A_{ext} = \prod D^2/4$$

Select the standard size cylinder with a diameter equal to or greater than the calculated value from the data on the manufacturer's domain. If required, modify the piston area (A_{ext}) and reselect the pump flow rate (Q_{Ap}) and check and revise the values as per the relations given in the above paragraphs.

Also, select the required piston-rod diameter from the manufacturer's datasheet.

Next, find the pressure (P) required in the hydraulic line to develop the necessary force from the following equation:

$$F = P \times A$$

The working pressure can be calculated by taking into account the pressure drops (about 15%) in the hydraulic power transmission system. Check the hydraulic power (P_{hyd}) and reconcile. This pressure should be less than the maximum pressure rating of the pump or any other component that is selected. Remember, this pressure can be set by using a pressure relief valve.

Next, calculate the theoretical pump flow rate (Q_{TP}) from the following equation:

$$Q_{TP} = Q_A/ \eta_{vP}$$

Find the volumetric displacement of the pump from the following equation:

$$V_{DP} = Q_{TP}/N_p$$

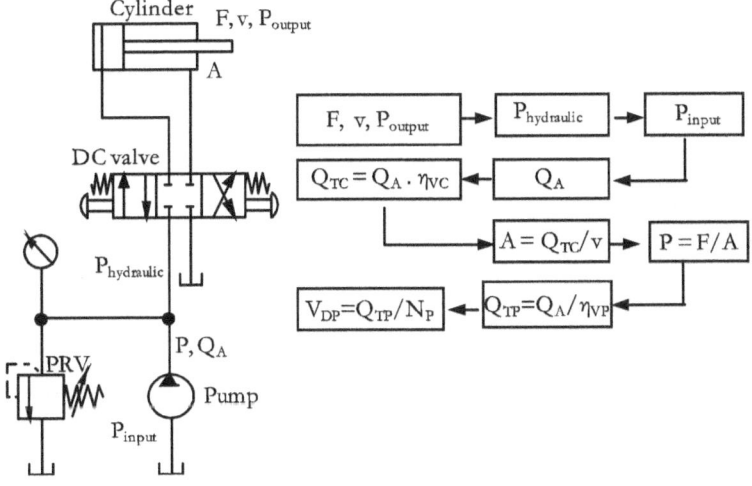

Figure 14.1 | An approach to hydraulic design

Figure 14.1 shows the basic steps, as used in this book, to design hydraulic systems.

All other parameters of the system can be calculated or selected using steps as given in the Design Example 1.

Design Example 1

Design the hydraulically-driven, 2200 pound device to meet a positive load in an industrial working environment. The device involves the following medium quality components: (1) Double-acting cylinder with a stroke of 50 cm, (2) Pump and coupled motor, (3) Reservoir, (4) Fluid, (5) Strainer and Filters, (6) PRV, (7), Pressure gauge, (8) 4/3-DC valve, and necessary conductors.

The cylinder is to move with a maximum speed of 0.328 ft/s. Assume the ambient temperature to be 85°F and the maximum design temperature as 160°F. Also, draw the hydraulic circuit to implement the control scheme. The leakage flows in the pump and actuator can be taken at least 10% (each), and frictional losses in the pump and actuator can be taken at least 10% (each). The leakage flows in the pump and actuator can be taken at least 10% (each), and frictional losses in the pump and actuator can be taken at least 10% (each). (Note: The design of the conductor system is given in Design Example 2.

Solution
Control Circuit

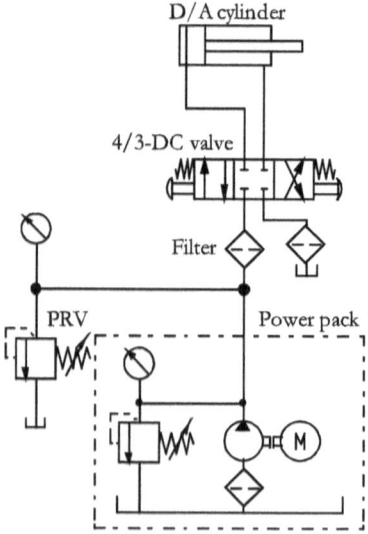

Figure 14.2 | Control circuit for Design Example 1

Design Calculations Part 1 [Design Example 1]

Force, F (Given)	= 2200 lb	
Velocity, v (Given)	= 0.328 ft/s	
Power rating, output	= 1.312 hp	(Power = F v/550 hp)
Stroke, S (Given)	= 1.6 ft	
Type of hydraulic system	Conventional	Conventional / Hydrostatic
Type of displacement	Constant	Constant / Variable
Leakage	= 10%	Each of pump & actuator
Frictional loss	= 10%	Each of pump & actuator
Volumetric efficiency, Actuator, η_{vC}	= 90% = 0.90	
Mechanical efficiency, Actuator, η_{mC}	= 90% = 0.90	
Overall efficiency, Actuator, η_{oC}	= 81% = 0.81	
Volumetric efficiency, Pump, η_{vP}	= 90% = 0.90	
Mechanical efficiency, Pump, η_{mP}	= 90% = 0.90	
Overall efficiency, Pump, η_{oP}	= 81% = 0.81	
Duty cycle	= 100%	Uninterrupted / With pause x sec
Type of load	Positive load	Positive Load / Negative Load (Use counter-balance valve, Pilot-operated check valve) / Inertia Load
Presence of shock	Negligible	Negligible / Moderate / Severe
Shock absorption	Not required	Required (Use accumulator) / Not required
Speed control	Not required	Meter-in / Meter-out / Bleed-off / Regeneration
No. of actuators	One	Single / Multiple
Energy-saving feature	Not required	Required (Use tandem center valve | Unloading valve) / Not required

Energy storage	Not required	Required (Use accumulator) / Not required
Leakage compensation	Not required	Required (Use accumulator) / Not required
Application environment	Normal	Normal / hot / environmentally-sensitive / Hygiene / hazardous
System quality requirement	Medium	Low / Medium / High
Ambient temperature	= 85°F	
Maximum system temperature	= 160°F	
Power Requirements		
Power, output	= 1.312 hp	(Power = F v/550 hp)
Power, hydraulic ($P_{hyd} = P_{out} / \eta_{oC}$)	= 1.312/0.81 = 1.62 hp	
Power, input / Electric motor ($P_{input} = P_{hyd} / \eta_{oP}$)	= 1.62/0.81 = 2 hp (single phase power supply)	Gear (hp): ¼, 1/3, ½, ¾, 1, 1½, **2**, 3, 5 Vane (hp): 7, 13.6, 17.8, 20.4, 22.8, 24, 24.5, 29.5, 32.5, 35.5, 36, 37.5, 39 Piston (hp): 16.3, 17.7, 20.2, 35, 35.5, 64, 73, 74, 80, 89, 95, 100, 150, 215
Cylinder		
Flow rate, Q_A (From Manufacturer's datasheet)	= 3.4 gpm	The hydraulic power requirement has to be met with proper values of P and Q_A selected from the data on OEM. Once QA is fixed, the size of all other components can be decided.
Flow rate, Q_{TC} (= Q_A x η_{VC}) 1 gpm = 0.002228 ft³/s	= 3.4 x 0.9 = 3.06 gpm =0.00682 ft³/s	
Area of cylinder, preliminary, A_{ext} ($Q_{TC} = A_{ext}$ x v)	= 0.00682 /0.328 = 0.0208 ft² = 3 in²	1 ft² = 144 in²
Diameter of the cylinder, D ($A = \prod D^2/4$)	= 1.95 in	

Diameter of the cylinder, D (Select standard values)	= 2 in	1, 1.5, 2, 2.5, 3.25, 4, 5, 6, 8
Area of the cylinder, revised, (A_{ext})	= 3.14 in²	
Piston Rod dia (Select from manufacturer's data sheet)	= 0.625 in	Piston Dia (Rod dia): 1 (0.5, 0.625), 1.5 (0.625, 1), 2 (0.625, 1, 1.375), 2.5 (0.625, 1, 1.75, 1.375), 3.25 (1, 1.75, 2, 1.375), 4 (1, 1.375, 2.5, 1.75, 2), 5 (1, 1.75, 3.5, 2, 2.5, 3), 6 (1.75, 4), 8 (2, 5.5)
Pressure, preliminary, P (P = F/A)	= 2200/3.14 = 700 psi	
Pressure drop, expected, ΔP	= 105 psi	Assumed: 15% of P
Set pressure, PRV	= 805 psi	
Set pressure, PRV, rounded off	= 810 psi	Pressure setting of the PRV; The pressure ratings of all system components should be higher than the pressure setting of the PRV.
Check, Power hydraulic (=P(psi)xQ_A(gpm)/1714 hp)	= 810x3.4/1714 = 1.61 hp	1.61 hp ~ 1.62 hp (OK)
Cylinder type	Double-acting (Tie-rod type)	Tie-rod, mill type, welded, threaded
Shock cushioning	Not required	
Seal material	NBR	NBR/Viton/Teflon
Mounting, cylinder body	Tie-rod mounting	Tie-rod, Flange, Foot, Trunnion
Mounting, piston-rod	Piston-rod end threads	Piston-rod end threads, rod clevis, eye bracket, knuckle, pivot pin, pin-hole

Power Pack (Pump-Motor unit)		
Pump flow rate, theoretical, Q_{TP} 1 gpm = 0.002228 ft³/s 1 ft³/s = 1728 in³/s	= 3.4/0.9 gpm = 3.78 gpm = 0.00842 ft³/s = 14.55 in³/s	$\eta_{vP}=Q_A/Q_{TP}$ [Flow rate, which is necessary to ensure the required velocities or revolutions]
Speed, pump drive shaft, N (Selection)	= 1725 rpm	Select from the data in the manufacturer's domain
Speed, pump drive, n	= 28.75 rps	
Pump displacement, V_D (= Q_{TP}/n)	= 14.55/28.75 = 0.5 in³/rev	
Select available displacement from catalogue	= 0.455304 in³/rev	
Type of pump	Gear	Gear / Vane / Piston
Theoretical torque, T_T [= V_D (in³/rev)x P (psi) /(2∏]	= 0.455304 x 810/ 2∏ = 58.73 in.lb	
Actual torque, T_A (= P_{input} x 63025/N)	= 2 x 63025/1725 = 73.07 in.lb	
Tank size multiplying factor	= 3	
Reservoir oil capacity	= 3 x 3.4 gallons = 10.2 gallons	10, 20, 30, 60, 100, 150 gal
Tank size (Selection)	= 10 gal	10, 20, 30, 60, 100, 150 gal
Tank dimension (L)	= 22 in	10 gal – 22x18x20 in³ 20 gal – 30x18x20x in³ 30 gal – 36x24x20 in³ 40 gal – 36x24x22 in³ 50 gal – 36x24x23 in³ 60 gal – 48x27x21 in³ 120 gal – 60x30x25 in³
Tank dimension (W)	= 18 in	
Tank dimension (H)	= 20 in	
Tank Surface area (=2xLxW+2xWxD+2xDxL)	= 2392 in² =16.6 ft²	ft² = in²/144
Tank volume (=LxWxD)	= 7920 in³ = 34.29 gallons	Gallons = in³/231
Oil height in the tank, for 10 gallons of oil	=10x231/(22x18) =5.83 in	
The distance between the oil level and the centre of pump shaft	=20 – 5.83 =14.17 in	<36 in (3 ft) OK

Tank thickness	3/16" (0.1875")	Up to 20 gallons – 12 to 14 gauge (0.1" – 0.078") From 20 to 100 gallons – 3/16" >100 gallons – 3/8"
Top cover thickness	3/16" (0.1875")	Up to 50 gallons – 3/16" 50 to 300 gallons – 3/8" >300 gallons – ½"
Tank and top cover material	Steel	Hot-rolled steel/Aluminium
Clean-out plate	Required	
Cleanout cover, size	10" dia	Up to 35 gallons – 10" Above 50 gallons – 14"
Cleanout cover material	Steel	
Gaskets for cleanout cover	Required	
Surface treatment	Zinc-coated	With or without Zinc-coated /
Jointing method for Suction/Return pipe with top plate	Rubber grommet	Rubber grommet for indoor applications / Welded couplings for outdoor use
Baffle with cutouts	Required	Cutout for oil circulation
Layout of the drive	Vertical on the tank	Vertical on the tank/ other
Heat load estimate (Assumed: 50% of motor hp)	= 2 x 0.5 hp = 1 hp	
Heat dissipation rate by reservoir, H(kW)	= 0.001x(160-85)x16.6 = 1.245 hp	H (hp) = 0.001 x ΔT (^{0}F) x A (ft^2)
Heat exchanger	Not required	
Heat exchanger capacity	= 0	
Oil flow rate, heat exchanger	-	
Fluid level indicator	Required	
Fluid level switch	Not required	
Thermometer/ Thermostat	Required (Thermometer)	
Magnet	Required	
Drain plug	Required	
Sound level	<75dB(A)@1m	

Fluid		
Type	Mineral-based	Mineral / fire-resistant /bio-degradable/ food-g
Max. fluid temp, t_{op}	160°F	
Viscosity, ISO VG	46 (or 200 SSU@100°F)	16 - 36 cSt at t_{op}
Flow, maximum	3.4 gpm	
Viscosity index	90	90 to 110
Fluid cleanliness level (As per ISO 4406)	18/16/13	NAS Code 7
Filters		
suction strainer	149 μ	149μ
Suction filter	90 μ (Paper filer)	90μ
Return-line filter	25 μ (Nominal) paper filter with built-in by-pass (setting 14.5 psid), and visual indicator/ electrical indicator	Beta ratio $\beta_{25©} \geq 100$ [With synthetic fluid 10 μ or better]
Pressure filter	10 μ (absolute) glass fibre filter with built-in by-pass (35 psid), visual indicator/electrical indicator	14/12/9 & 15/13/10 ($\beta_{10©} \geq 100$ ($\geq 99\%$) 16/14/11 & 17/15/12 ($\beta7© \geq 100$) 18/16/13 ($\beta10© \geq 100$) 19/17/14 ($\beta15© \geq 100$) 20/18/15 ($\beta20© \geq 100$) Micron rating: 3μ (absolute), 6μ (absolute), 10μ (absolute), / 25 μ (Nominal)/ $\Delta P=35$ psid
Off-line filter	Not required	
Air filter	5μ	3μ, 5μ, 10μ
Gauge	2.5 in, without gauge isolator	Standard 2.5 in liquid-filled pressure gauge with or without isolator

Pressure Relief Valve		
Type	Pilot-operated	Direct-acting or pilot-operated
Flow capacity	> 3.4 gpm	
Size	NFPA D03/ CETOP03	NG6 (D03) NG10 (DO5) NG16 (DO7) NG25 (DO8) NG32 (DO10) The NG6 valve can provide high functional limits up to 20 gpm in combination with an energy-saving, very low pressure drop.
Pressure rating	> 810 psi	
Protection	-	IP 55

Directional Control Valve		
Type	4/3, Closed center	
Port size	NFPA D03	NG6 (D03) NG10 (DO5) NG16 (DO7) NG25 (DO8) NG32 (DO10) The DO3 valve can provide high functional limits up to 20 gpm in combination with an energy-saving, low pressure drop.
Pressure rating	> 810 psi	
Actuation	Manual	Manual or Solenoid-operated
Seals	NBR	NBR/ FPM/ PTFE
Mounting Interface	NFPA	DIN 24340 A6 / ISO 4401 / CETOP RP 121-H / NFPA D03
Control voltage	Not applicable	Select the standard control voltage of one's region

Summary of Steps for the Pump-Cylinder Design

Through system analysis, find the following parameters of the hydraulic system: F, v, cylinder efficiencies, pump efficiencies, design temperature, ambient temperature, etc.

Determination of Power Ratings
1. Find output power, $P_{out} = F \cdot v$ (in hp)
2. Find hydraulic power, $P_{hyd} = P_{out}/\eta_{oC}$ (in hp)
3. Find input (electric motor) $P_{input} = P_{hyd}/\eta_{oP}$ (in hp)

Determination of Flow Rate
4. Select Q_A value from pump data on manufacturer's domain based on the rating of the electric motor

Cylinder Design
5. Find Q_{Tc} of the cylinder ($Q_{TC} = \eta_{vC} \cdot Q_A$)
6. Find area A of the cylinder ($A = Q_{TC}/v$)
7. Find the bore diameter of the cylinder $[D = \sqrt{(4 \cdot A /\pi)}]$
8. Select the standard bore diameter (D) of the cylinder from the cylinder data on manufacturer's domain
9. Select the rod diameter from the cylinder data on manufacturer's domain

Determination of Pressure
10. Find the preliminary value of pressure ($P_{preliminary} = F/A$)
11. Set the pressure (P) about 10 to 15% more than $P_{priliminary}$
12. Calculate P_{hyd} from P and Q_A values
    $[P_{hyd} (hp) = P(psi) \cdot Q_A (gpm)/1714]$
13. Compare the value at step 12 with that at step 2, and reconcile

Design of Pump

14. Find the theoretical flow rate, $Q_{TP} = Q_A / \eta_{vP}$
15. Select the speed of the drive shaft, N (rpm) or n (rps)
16. Find the pump displacement in cc, $V_D = Q_{TP}/N$
    $[V_D \text{ (in}^3/\text{rev)} = Q_{TP} \text{ (in}^3/\text{s)} / \text{ n (rps)}]$
17. Select the nearest pump displacement (V_D) of the pump from the pump data on manufacturer's domain
18. Find the actual torque, T_A
    $[P_{input} \text{ (hp)} = T_A \text{ (in.lb)} . \text{ N (rpm)}/63025]$
19. Find the theoretical torque, T_T
    $[T_T = V_D \text{ (in}^3/\text{rev)} . \text{ P (psi)} / 2\pi]$
20. Reconcile the torque values

Hydraulic Fluid Selection

21. Select the fluid type (Mineral, fire-resistant water-based, fire-resistant synthetic, bio-degradable, food-grade)
22. Select the fluid viscosity
23. Select the fluid VI (90 to 110)
24. Set the fluid cleanliness level (As per the SAE/ISO standard)

Reservoir Design

25. Determine the fluid volume = 3 to 5 times Q_{TP}
26. Select tank size (L, W, H) from the data on manufacturer's domain
27. Decide the thicknesses of the tank and top plates
28. Determine the size of the baffle plate
29. Determine the surface area ($A_{surface}$) of the tank
30. Determine the volume of the tank
31. Find the heat load (= 30 to 50% of the motor rating)
32. Find the heat dissipation rate $[=0.001 . \Delta T \text{ (°F)} . A_{surface}(\text{ft}^2)]$
33. Determine the necessity of heat exchanger and its rating

Specifications of Other Components

34. Determine the specifications of filters (Strainer, suction filter, pressure filter, return-line filter, off-line filter)
35. Determine the specifications of the PRV
36. Determine the specifications of the directional control valve

Design Example 2

Design a fluid conductor system for the problem in Design Example 1.

Solution

Suction Line		
Viscosity at operating temperature (Assumed)	= 30 cSt	$1\ cSt = 10^{-6}\ m^2/s$
Fluid velocity (Suction side), $v_{suction}$	= 4 ft/s	Velocity Vs Viscosity 4 ft/s - 30 cSt 3.6 ft/s - 50 cSt 2.5 ft/s - 100 cSt 2 ft/s - 150 cSt
Area of suction pipe/tube ($Q_{TP}/v_{suction}$)	$= 0.00842/4\ ft^2$ $= 1.212\ in^2$	$1\ ft^2 = 144\ in^2$
Internal dia of the pipe/tube, $D_{i/suction}$	$= \sqrt{(4 \times 1.212/\pi)}$ = 1.24 in	
Type of conductor (Pipe/Tube/Hose)	Tube	Tubing can be considered ($\leq 1\frac{1}{4}$ in)
Outside dia of tube ($D_{o/suction}$)	= 1.5 in	Tubing sizes (in): 3/16, 1/4, 5/16, 3/8, 1/2, 5/8, 3/4, 1, 1¼, 1½
Wall thickness of the tubing ($t_{suction}$)	= 0.12 in	OD (Wall thickness): 3/16 (0.035), 1/4 (0.035, 0.049), 5/16 (0.049, 0.065), 3/8 (0.049, 0.065), 1/2 (0.049, 0.065, 0.083), 5/8 (0.065, 0.095), 3/4 (0.065, 0.095, 0.109), 1 (0.065, 0.095, 0.120), 1¼ (0.095, 0.120), 1½ (0.12, 0.134)
Internal dia of the tube, $D_{i/suction}$ (Revised) (Outside dia - 2 x Wall thickness)	= 1.26 in	Reconcile
Reynolds Number, $R_{e,Suction\ side}$ [$=7740 \times v(ft/s) \times d(in)/\upsilon(cSt)$]	$= 7740 \times 4 \times 1.26/30$ = 1300 (OK)	Is it less than 2000?
Type of flow	Laminar	Laminar / Turbulent
Tubing material	Seamless cold-drawn carbon steel	Seamless cold-drawn carbon steel AISI 4130 steel

Pressure Line		
System pressure, Final	= 810 psi	See Page 83
Flow rate (Q_A)	= 3.4 gpm = 0.007575 ft^3/s	See page 82 1 gpm = 0.002228 ft^3/s
Fluid velocity (pressure line), $v_{pressure-line}$	= 13 ft/s	Pressure -Velocity: psi (ft/s): 900 – 1450 (13 – 14) 1450 – 2300 (14 – 16) 2300 – 3600 (16 – 18) 3600 – 5800 (18-20)
Area of the pressure line conductor	= 0.007575/13 ft^2 = 0.000583 ft^2 =0.084 in^2	[A =Q_A (ft^3/s) /v (ft/s)] 1 ft^2 = 144 in^2
Internal dia of pressure line conductor, $D_{i,pressure-line}$	= $\sqrt{(4\times0.084/\pi)}$ = 0.327 inch	
Type of conductor (Pipe/Tube/Hose)	Tube	Tubing can be considered ($\leq$1¼ in)
Outside dia of tube $(D_{o/pressure\_line})$	= ½ in = 0.5 in	Tubing sizes (in): 3/16, ¼, 5/16, 3/8, ½, 5/8, ¾, 1, 1¼, 1½
Wall thickness of the tubing $(W_{pressure\_line})$	= 0.049 in	OD (Wall thickness): 3/16 (0.035), ¼ (0.035, 0.049), 5/16 (0.049, 0.065), 3/8 (0.049, 0.065), ½ (0.049, 0.065, 0.083), 5/8 (0.065, 0.095), ¾ (0.065, 0.095, 0.109), 1 (0.065, 0.095, 0.120), 1¼ (0.095, 0.120), 1½ (0.12, 0.134)
Internal dia of the tube, $D_{i/pressure\_line}$ (Revised)	= 0.5–(2x0.049) in = 0.402 in	(Outside dia – 2 x Wall thickness)Reconcile/OK
Reynolds Number, $R_{e,pressure\_line}$ [=7740xv(ft/s)xd(in)/v(cSt)]	= 7740x13x0.402/30 = 1348 (OK)	Is it less than 2000?
Type of flow	Laminar	Laminar / Turbulent
Tubing material	Seamless cold-drawn carbon steel	Seamless cold-drawn carbon steel / AISI 4130 steel
Pipe tensile strength, S	= 55000 psi	Tensile strength: Cold-drawn carbon steel - 55000 psi AISI 4130 steel -75000 psi
Length, pressure line, L	= 6.5 ft	(Assumed)

ID of the tube, D	$= 0.115$ in $= 0.00958$ ft	
Fluid density, ϱ	$= 1.6$ slug/ft^3	
Fluid velocity, v	$= 13$ ft/s	
Reynolds number, R_e	$= 1348$	See Page 91
Friction factor , f	$= 64/1348$ $= 0.0475$	64 / Re
Pressure loss, ΔP_l 0.0475x1.6x6.5x13^2/2x0.00958	$= 4357$ lb/ft^2 $= 30$ psi	$\Delta P = f . (\varrho L/2D) . v^2$
No. of 90° elbows	2	
No. of T fittings	2	
Pressure drop across two 90^0 elbows, ΔP_e	$= \zeta \frac{\varrho .v^2}{2}$ x 2 $= 0.2$ x1.6x13^2 x2/2 $= 54$ lb/ft^2 $= 0.375$ psi	[Loss coefficient, ζ] 90^0 elbow – 0.2 45^0 elbow- 0.15 Tee fitting – 0.9
Pressure drop across two T fittings, ΔP_t	$= \zeta \frac{\varrho .v^2}{2}$ x 2 $= 0.9$ x1.6 x13^2 $= 243$ lb/ft^2 $= 1.69$ psi	Sharp entrance - 0.5 Rounded entrance - 0.05 Sharp edged exit - 1 Rounded exit - 1
Total pressure drop	$= \Delta P_l + \Delta P_e + \Delta P_t$ $= 30+0.375+1.69$ psi $= 32$ psi [3.9% of P]	Should be within 3 to 5% of the system pressure
Factor of Safety, SF	$= 8$	<Shock & severe mechanical strain - 8, Considerable shock & mechanical strain - 6, Mechanical strain not considerable - 4
Burst pressure, BP [=2tS/ (Di)]	=2x0.049x55000/0.402 $= 13408$ psi	
Working pressure (=BP/SF)	$= 13408/8$ psi $= 1676$ psi >810 psi (OK)	Reconcile / The selected wall thickness is acceptable or not.
Wall thickness	$= 0.049$ in (OK)	OK / Not OK
Hose, type	Teflon-lined, steel braid Standard length	
Hose length	As required	

Return Line		
Flow rate (Q_A)	= 3.4 gpm = 0.007575 ft³/s	
Fluid velocity (return line), $v_{return-line}$	= 6.5 ft/s	Velocity: 6.5 to 10 ft/s
Area of return line conductor	= 0.007575/6.5 ft² = 0.001156 ft² = 0.166534 in²	[A =Q_A (ft³/s) /v (ft/s)] 1 ft² = 144 in²
Internal dia of the return line conductor, $D_{i,return-line}$	= √(4x0.166532/π) = 0.3855 in	
Type of conductor (Pipe/Tube/Hose)	Tube	Tubing can be considered (≤1¼ in)
Outside dia of tube ($D_{o/return\_line}$)	= 0.5 in	Tubing sizes (in): 3/16, ¼, 5/16, 3/8, ½, 5/8, ¾, 1, 1¼, 1½
Wall thickness of the tubing (W_{return_line})	= 0.049 in	OD (Wall thickness): 3/16 (0.035), ¼ (0.035, 0.049), 5/16 (0.049, 0.065), 3/8 (0.049, 0.065), ½ (0.049, 0.065, 0.083), 5/8 (0.065, 0.095), ¾ (0.065, 0.095, 0.109), 1 (0.065, 0.095, 0.120), 1¼ (0.095, 0.120),
Internal dia of the tube, $D_{i/return}$ (Revised)	= 0.402 in	(Outside dia - 2xwall thickness)Reconcile/OK
Reynolds Number, $R_{e,return\_line}$	= 7740x6.5x0.402/30 = 674 (OK)	[=7740xv(ft/s)xd(in)/υ(cSt)]
Type of flow	Laminar	Laminar / Turbulent
Tubing material	Seamless cold-drawn carbon steel	Seamless cold-drawn carbon steel	AISI 4130 steel	
Tubing Joint	37° flared fitting	<1000 psi - 37⁰ flared fitting	1000 to 3000 psi - 45⁰ flared fitting	 For medium / heavy wall tubing – Flare-less fitting
Pipe/Tubing Supports	As required	Plain mounting or damped mounting

Summary of Steps for the Design of a conductor system

Suction line
1. Determine the fluid flow rate, Q_{TP}, Q_{AP}
2. Assume fluid velocity (suction side), $v_{suction}$
3. Find the area of suction pipe/tube $(Q_{TP}/v_{suction})$
4. Determine the initial value of the internal dia, $D_{i/suction}$
5. Select the type of conductor (Pipe/Tube/Hose)
6. Select the outside dia of the conductor, $(D_{o/suction})$
7. Select the wall thickness of the conductor, $t_{suction}$
8. Revise the final value of the internal dia, $D_{i/suction}$
9. Calculate the Reynolds Number, R_e $[=7740xv(ft/s)xd(in)/\upsilon(cSt)]$
10. Ensure laminar flow

Pressure line
11. Assume fluid velocity (pressure line), $v_{pressure}$
12. Find the area of pressure line pipe/tube $(Q_{AP}/v_{pressure})$
13. Determine the initial value of the internal dia, $D_{i/pressure}$
14. Select the type of conductor (Pipe/Tube/Hose)
15. Select the outside dia of the conductor, $(D_{o/pressure})$
16. Select the wall thickness of the conductor, $t_{pressure}$
17. Revise the final value of the internal dia, $D_{i/pressure}$
18. Calculate the Reynolds Number and ensure laminar flow
19. Determine the burst pressure, BP $[=2tS/ (Di)]$
20. Determine the working pressure $(=BP/SF)$
21. Ensure sufficient wall thickness

Return line
22. Assume fluid velocity (return side), v_{return}
23. Find the area of return pipe/tube $(\sim Q_A/v_{return})$
24. Determine the initial value of the internal dia, $D_{i/return}$
25. Select the type of conductor (Pipe/Tube/Hose)
26. Select the outside dia of the conductor, $(D_{o/return})$
27. Select the wall thickness of the conductor, t_{return}
28. Revise the final value of the internal dia, $D_{i/return}$
29. Calculate the Reynolds Number and ensure laminar flow
30. Select the type of material for suction, pressure, return line

Design Example 3

Design a 4.7 hp hydraulic system for mixing operation using a hydraulic motor. The system involves the following elements: (1) hydraulic motor, (2) pump and coupled electric motor, (3) reservoir, (4) strainer, (5) PRV, (6) 4/3-DC float-center valve, and (9) necessary tubing.

The motor torque requirement is about 342 lb.in. The leakage flows can be up to a maximum of 5 to 10% each for the pump and hydraulic motor. Mechanical losses can be assumed to be 10% each for the pump and hydraulic motor. Assume the ambient temperature as 85°F and the maximum design temperature as 150°F. The return pressure is about 90 psi. Also, draw the hydraulic circuit to implement the scheme.

Solution

Control Circuit

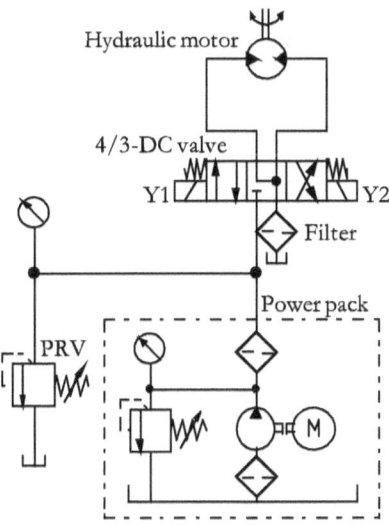

Figure 14.3 | Control circuit for Design Example 3

Power rating	4.7 hp	Given
Torque requirement, T_{AM}	342 lb.in	Given
Leakage, pump	5%	Given
Leakage, motor	10%	Given
Mechanical losses, pump	10%	Given
Mechanical losses, motor	10%	Given
Volumetric efficiency, motor, η_{vM}	90% = 0.90	
Mechanical efficiency, motor, η_{mM}	90% = 0.90	
Overall efficiency, motor, η_{oM}	81% = 0.81	
Volumetric efficiency, Pump, η_{vP}	95% = 0.95	
Mechanical efficiency, Pump, η_{mP}	90% = 0.90	
Overall efficiency, Pump, η_{oP}	81% = 0.855	
Duty cycle	100%	Given
Return pressure (Given)	90 psi	Given
Type, hydraulic system	Conventional	Conventional/ Hydrostatic
Ambient temperature	85°F	Given
Maximum system temperature	150°F	Given
Application environment	Normal	Normal/hot/ Hygiene-specific/environmentally -sensitive
Power Requirements		
Power, output, P_{out}	4.7 hp	
Power, hydraulic	=4.7/0.81 = 5.8 hp	$(P_{hyd} = P_{out} / \eta_{oM})$
Power, input / Electric Motor	=5.8/0.855 = 6.78 hp	$(P_{input} = P_{hyd} / \eta_{oP})$

Determination of Flow rate, actual		
Volumetric disp. (V_{DP}), pump (Selection)	$=1.22$ in^3/rev	From OEM pump data [Page 112]
Drive speed	$=1455$ rpm $=24.25$ rps	
Flow rate, theoretical, $Q_{TP} = V_{DP}$ x n	$=29.585$ in^3/s $=0.01712$ ft^3/s $=7.68$ gpm	$=1.22$ x 24.25 in^3/s 1ft^3/s = 1728 in^3/s 1 gpm=0.0022288 ft^3/s
Flow rate, actual, $Q_{AP} = Q_{TP}$ x η_{vP}	$=28.11$ in^3/s $=0.016264$ ft^3/s $=7.296$ gpm	$=29.585$ x0.95 in^3/s $=0.01712$x0.95 ft^3/s $=7.68$x0.95 gpm

Hydraulic Motor		
Torque requirement, T_{AM}	$= 342$ lb.in	Given
Speed max, Motor, N_M	$= 4.7$ x $63025/342$ $= 866$ rpm $= 14.43$ rps	P_{out}(hp)=T_{AM} (lb.in)x N_M(rpm) /63025
Actual flow rate, Q_A	$=28.11$ in^3/s	
Theoretical flow rate, Motor, Q_{TM} $(= Q_A$ x $\eta_{VM})$	$= 28.11$ x 0.9 $= 25.299$ in^3/s	$= 0.0146$ ft^3/s $= 6.6$ gpm
Volumetric displacement, motor, V_{DM} [Calculation]	$= 25.299/14.43$ $= 1.7532$ in^3/rev	$Q_{TM} = V_{DM}$ x N_M
Hydraulic motor displacement, V_{DM}[Select from manufacturer's domain]	$= 1.93$ in^3/rev [From Table A2.4, page 120]	Typical values: 0.5, 0.79, 1.21, **1.93**, 2.2, 2.8, 3, 3.6, 4.5, 5.9, 7.3, 8.9, 9.7, 10, 11.3, 14.1, 17.9, 22.6, 24, 28.7, 29.2, 30, 33, 34.9, 38, 39.6, 40.6, 41.7, 46, 49, 57.4, 60, 76, 95 in^3/r
Theoretical torque, T_T $(=T_{AM}/\eta_{mM})$	$= 342/0.9$ $= 380$ lb.in	$T_T = T_{AM} / \eta_{mM}$
Pressure differential, (ΔP)	$= 380$x2x$\pi/1.93$ $= 1236$ psi	$(T_T=V_{DM}$ x $\Delta P/2\pi)$
Return pressure	$= 90$ psi	Given
Pressure, P, preliminary	$= 1236 + 90$ $= 1326$ psi	$P_{pre} = \Delta P + P_{return}$
Set pressure, PRV	$= 1326$ x 1.1 ~ 1450 psi	Assumed: 10% of P

Pressure rating of the hydraulic motor	= 1500 psi	Should be greater than the set pressure
Check, power, hydraulic	= 1450x7.3/1714 = 6.17 hp > 5.8 hp (OK)	Power =P(psi)xQ_A(gpm)/1714
Motor type	LSHT	LSHT | HSLT
Seal material	NBR	NBR/Viton/Teflon
Mounting, Motor body	Flange	Flange (SAE/Metric) / Foot
Mounting, Shaft	Straight-keyed Shaft	Shaft (Straight-keyed, tapered, splined, threaded)
Case drain	Required	

Power Pack (Pump-Motor unit)		
Power, input / Electric Motor	=7.2 hp =7.4 hp (selection)	
Volumetric disp. (V_{DP}), pump (Selection)	=1.22 in^3/rev	From OEM pump data [Page 112]
Drive speed	=1455 rpm =24.25 rps	0.55 hp (1395, 2790), 0.75 hp (1395, 2850), 1.1 hp (1410, 2835), 1.5 hp (1410, 2860), 2.2 hp (1420, 2850), 3.0 hp (1420, 2895), 4.0 hp(1440), 5.5 hp (1455), 7.5 hp (1455)
Flow rate, theoretical, $Q_{TP} = V_{DP}$ x n	=29.585 in^3/s =0.01712 ft^3/s =7.68 gpm	=1.22 x 24.25in^3/s [1 gpm=0.0022288 ft^3/s] [1ft^3/s = 1728 in^3/s]
Flow rate, actual, $Q_{AP} = Q_{TP}$ x η_{vP}	=28.11 in^3/s =0.016264 ft^3/s =7.296 gpm	=29.585 x0.9 in^3/s =0.01712x0.9 ft^3/s =7.68x0.9 gpm
Pressure requirement	= 1450 psi	
Inlet pressure	-3 psi (Mineral oil)	(-1.5 to -5 psi for synthetic fluids/water-in-oil emulsions)
Type of pump	Gear	Gear / Vane / Piston
Theoretical torque, T_{TP}	= 1.22 x 1450 / 2π = 282 lb.in	(T_T= V_{DP} x P/(2Π)
Actual torque, T_{AP}	= 7.4x63025/1455 = 321 lb.in	($T_{AP} = P_{input}$ x 63025/N)

Tank size multiplying factor	= 3	
Reservoir oil capacity	= 3 x 7.68 = 23 gallons	3 Q
Tank size (Selection)	= 30 gallons	10, 20, 30, 60, 100, 150 gal
Tank dimension (L)	36 in	10 gal – 22x18x20 in³
Tank dimension (W)	24 in	20 gal – 30x18x20x in³ 30 gal – 36x24x20 in³
Tank dimension (H)	20 in	40 gal – 36x24x22 in³ 50 gal – 36x24x23 in³ 60 gal – 48x27x21 in³ 120 gal – 60x30x25 in³
Tank Surface area (=2xLxW+2xWxH+2xHxL)	= 4128 in² =28.67 ft²	ft² = in²/144
Tank volume (=LxWxD)	= 17280 in³ = 74.81 gallons	Gallons = in³/231
Tank material	Steel	
Layout of the drive	Vertical on the tank	Vertical on the tank/ other
Heat load estimate (Assumed: 40% of motor hp)	= 7.4 x 0.4 hp = 2.96 hp	
Heat dissipation rate by the reservoir	=0.001x(160-85) x 28.67 = 2.15 hp	H (hp)= 0.001 x ΔT (^{0}F) x A (ft²)
Heat exchanger	Required	
Heat exchanger capacity	> 0.81 hp	
Oil flow rate, heat exchanger	> 28.11 in³/s > 0.016 ft³/s > 7.3 gpm	
Fluid level indicator	Required	
Thermometer	Required	
Thermostat	Not required	
Clean-out plate	Required	
Magnet	Required	
Drain plug	Required	

Fluid		
Type	Mineral-based	Mineral oil, Fire-resistant (Phosphate ester)-based fluids, Water emulsions in oil, Water-glycol fluids

Max. fluid temperature, t_{op}	150°F	With mineral oil: -up to 150°F, Water-based fluids: up to 120°) [Consult OEM]
Viscosity, ISO VG	= 46	16 - 36 cSt at the operating temperature recommended
Flow max	= 7.6 gpm	
Viscosity index	= 90	80 to 110
Fluid cleanness level (As per SAE 4059)	SAE Class 4 [Equivalent to ISO 18/16/13]	(As per ISO 4406 or SAE 4059: Class 1 to 12)
Filters		
suction strainer	149 μ	Wire mesh
Suction filter	90 μ	Paper filer
Return-line filter	25 μ (Nominal)	Paper filter
Pressure filter	10 μ (absolute)	Glass fibre filter
Gauge	63 mm	
Pressure Relief Valve		
Type	Pilot-operated	Direct-acting or pilot-operated
Flow capacity	>7.3 gpm	
Size	NG06	
Pressure rating	≥1500 psi	
Protection	IP 55	IP 55
Directional Control Valve		
Type	4/3, Closed center	
Size	NFPA D03	
Pressure rating	≥1500 psi	
Actuation	Solenoid-operated	Manual or Solenoid-operated
Seals	NBR	NBR/ FPM/ PTFE
Mounting Interface	NFPA	DIN 24340 A6 / ISO 4401 / CETOP RP 121-H / NFPA D03
Control voltage	As appropriate	24 V DC / 110 VAC

Note: Part 2 [Design Example 3] for the conductor design is similar to Design Example 2.

Summary of Steps for the Pump-Motor Design

Through system analysis, find the following parameters of the system: T_{AM}, N_M, motor efficiencies, pump efficiencies, design temperature, ambient temperature, etc.

Determination of Power Ratings
1. Find output power, P_{out} (hp) $= [T_{AM}(\text{lb.in}) \times N_M(\text{rpm})]/63025$
2. Find hydraulic power, $P_{hyd} = P_{out}/\eta_{oM}$ (hp)
3. Find input (electric motor) $P_{input} = P_{hyd}/\eta_{oP}$ (hp)

Determination of Flow Rate
4. Select Q_A value from pump data on manufacturer's domain based on the rating of the electric motor

Hydraulic Motor Design
5. Find Q_{TM} of the motor ($Q_{TM} = \eta_{vM} \cdot Q_A$)
6. Find the motor displacement $Q_{TM} = V_{DM} \times N_M$
7. Select the standard displacement of the motor from the data on manufacturer's domain

Determination of Pressure
8. Find the theoretical torque, $T_{TM} = T_{AM} / \eta_{mM}$
9. Assume return pressure, P_{return}
10. Find the differential pressure, ΔP ($T_{TM} = V_{DM} \times \Delta P/2\Pi$)
11. Determine the preliminary pressure, $P_{priliminary} = \Delta P + P_{return}$
12. Set the pressure (P) about 10 to 15% more than $P_{priliminary}$
13. Calculate P_{hyd} from P and Q_A values
    $[P_{hyd}$ (hp) $= P(\text{psi}) \times Q_A(\text{gpm})/1714]$
14. Compare the value at step 13 with that at step 2, and reconcile

Design of Pump

15. Find the theoretical flow rate, $Q_{TP} = Q_A / \eta_{vP}$
16. Select the speed of the drive shaft, N (rpm) or n (rps)
17. Find the pump displacement in cc, $V_D = Q_{TP}/N$
    [V_D (in³/rev) = Q_{TP} (in³/s) / n (rps)]
18. Select the nearest pump displacement (V_D) of the pump from the pump data on manufacturer's domain
19. Find the actual torque, T_A
    [P_{input} (hp) = T_A (in.lb) . N (rpm)/63025]
20. Find the theoretical torque, T_T [$T_T = V_D$(in³/rev).P (psi)/2π]
21. Reconcile the torque values

Hydraulic Fluid Selection

22. Select the fluid type (Mineral, fire-resistant water-based, fire-resistant synthetic, bio-degradable, food-grade)
23. Select the fluid viscosity
24. Select the fluid VI (90 to 110)
25. Set the fluid cleanliness level (As per the SAE/ISO standard)

Reservoir Design

26. Determine the fluid volume = 3 to 5 times Q_{TP}
27. Select tank size (L, W, H) from the data on manufacturer's domain
28. Decide the thicknesses of the tank and top plates
29. Determine the size of the baffle plate
30. Determine the surface area ($A_{surface}$) of the tank
31. Determine the volume of the tank
32. Find the heat load (= 30 to 50% of the motor rating)
33. Find the heat dissipation rate [=0.001 . ΔT (°F) . $A_{surface}$(ft²)]
34. Determine the necessity of a heat exchanger and its rating

Specifications of Other Components

35. Determine the specifications of filters (Strainer, suction filter, pressure filter, return-line filter, off-line filter)
36. Determine the specifications of the PRV
37. Determine the specifications of the directional control valve
38. Design the conductor system

Design Example 4

For Design Example 1, it is now required to use an accumulator to supply about 15% of the flow requirement so that the size of the pump can be reduced. Design the accumulator part for the hydraulic system. Also, draw the hydraulic circuit to implement the scheme.

Solution

Control Circuit

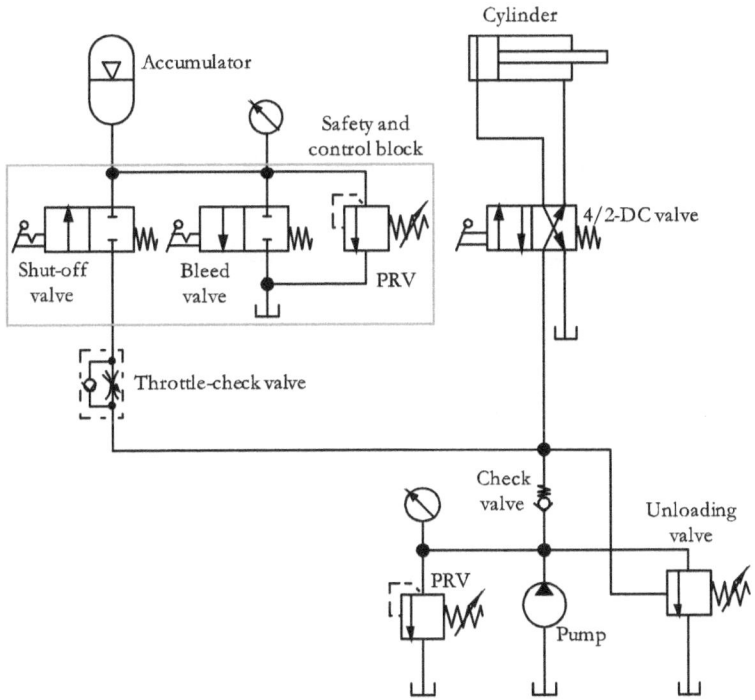

Figure 14.4 | Hydraulic circuit with accumulator

Preliminary data reproduced from Design Example 1		
Stroke, S (Given)	= 1.6 ft =19.2 in	
Energy storage	Required	
Flow rate, Q_A	= 3.4 gpm	
Cylinder Diameter, D	= 2 in	
Cylinder Area, (A_{ext})	= 3.14 in^2	
Set pressure, PRV	= 810 psi	
Pump displacement, V_D	=0.455304 in^3/rev	
Speed (Selection)	= 1725 rpm	
Accumulator Design		
Pump displacement, with accumulator, V_{DA}	=0.455304 in^3/rev	
Speed, pump drive shaft, N (Selection)	= 1725 rpm	
Pump flow rate, Q_A (with accumulator)	=0.382455 in^3/rev	84%
Accumulator flow rate	=0.072849 in^3/rev	16%
Cylinder volume, forward stroke	60.288 in^3	
Volume of fluid to be supplied by the accumulator, V	9.646 in^3	60.288 x 16%
Maximum System pressure, P_2	= 810 psi = 824.5 psi(a)	[Volume V_2]
Minimum system pressure, P_1	= 730 psi = 744.5 psi(a)	Assumed
Pre-charge pressure, P_0	= 730 x 80% = 584 psi = 598.5 psi(a)	Shock Absorber: 65-80% Energy storage – 80-90%
Accumulator Volume, $[V_0 = V/[(P_0/P_1) – (P_0/P_2)]$	=9.646/[(598.5/744.5)- 598.5/824.5)] = 123.67 in^3 = 0.53 gallons	A throttle valve should be incorporated into the accumulator to control the discharge flow rate.
Accumulator volume, (Selection)	=0.53 gallons	Diaphragm (gpm): 0.185, 0.37, 0.53, 0.74, 0.925 Bladder (gpm): 0.264, 0.66, 1, 2.64, 5.3, 9.25, 13.2
Type of accumulator	Diaphragm type	Diaphragm / Bladder / Piston
Working pressure	>810 psi	

Seal material	Buna-N	Buna-N, Butyl, Viton

Safety and control block		
Size of the shut-off valve	NG6	6, 10, 20, 32
Discharge	Manual bleed valve	Manual, 2-way solenoid (NO or NC)
Pressure relief valve	Required	PRV (3500 psi / 5000 psi)
Seal material	Buna-N	Buna-N, Butyl, Viton
Connection type	Threaded	Threaded (BSPP, SAE) Flanged (SAE)

Throttle check valve		
Rated flow	>3.4 gpm	
Maximum pressure	>810 psi	
Adjustment style	Screw	Screw/knob
Pressure Compensation	-	Pressure/temperature

Check valve		
Rated flow	>3.4 gpm	
Maximum pressure	>810 psi	

Unloading valve		
Rated flow	>3.4 gpm	
Maximum pressure	>810 psi	
Unloading pressure	Up to 1000 psi	1000 psi 2000 psi 3000 psi

Note: An accumulator which stores pressurized fluid can discharge its usable volume too rapidly as a downstream directional control valve is shifted. For this reason, the accumulator is equipped with flow control and bypass check at their inlet-outlet port to discharge the usable volume in the accumulator at a controlled rate.

Design Example 5 | Design of Hydraulic Systems with Multiple Actuators

A pump must provide sufficient power and flow required for all the actuators all the time for the smooth operation of the system. The preliminary step involved in designing a hydraulic system with multiple actuators is finding the power and flow rate requirements of every actuator in the system. Next, it is essential to find the sequence of operations of the actuators to find the maximum power and flow rate requirements of the actuators at any point in time. For an optimum design, the power and flow requirement must be just enough, not more, not less.

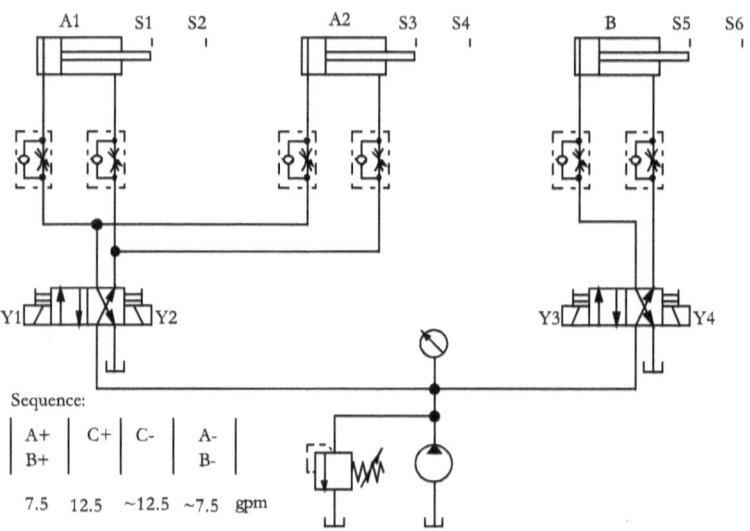

Figure 14.5 | A hydraulic system for clamping and punching operations

For example consider, a hydraulic system for clamping and punching operations. The system uses two identical cylinders (A and B) for the clamping operation, and a cylinder (C) for the punching operation. The circuit diagram of the system is given in Figure 14.5.

Next, the sequence of operation of actuators is determined and is given below:

Sequence:

$$\begin{array}{|c|c|c|c|}
\text{A+} & \text{C+} & \text{C-} & \text{A-} \\
\text{B+} & & & \text{B-}
\end{array}$$

It may be noted that the forward and return motions of the cylinders A and B are simultaneous. A sample requirements of force, speed, and power for every actuator are given in Table 14.1.

Table 14.1

Cylinder	Force (N)	Speed (m/s)	Power (Watt)	Flow rate (lpm)
A	10000	0.1	1000	14
B	10000	0.1	1000	14
C	30000	0.3	9000	47

From the analysis of the information derived so far, it can be observed that the maximum power requirement at any point in time is 9000 watt and the maximum flow rate requirement at any one point in time is 47 lpm.

There are a number approaches to finding a solution. Some of the alternative approaches are highlighted below.

(1) A single pump can be selected to provide the flow and power requirement of the system. The system can be set for the maximum of the pressure requirements of the actuators. If there are wide variations in the pressure requirements of actuators, then pressure reducing valves can be used to provide reduced pressures to some parts of the circuit.
(2) Multiple pumps can be used to meet the power, flow, and pressure requirements of the system.

An assessment may be carried out to select a design approach that would result in high efficiency and low cost.

Appendix 1

Pump Data

Appendix 1 gives the typical pump data extracted from reputed manufacturer's datasheet.

Pump Data – External Gear Pumps (with 3-phase Motor)

Motor (hp)	Pump Type	Speed (rpm)	Displacement, in³/rev	Flow rate (gpm)	Pressure, continuous (psi)	Phase
¼	Gear	1725	0.053565	0.4	928	3
¼	Gear	1725	0.080348	0.6	638	3
¼	Gear	1725	0.093739	0.7	479	3
¼	Gear	1725	0.120522	0.9	377	3

Motor (hp)	Pump Type	Speed (rpm)	Displacement, in³/rev	Flow rate (gpm)	Pressure, continuous (psi)	Phase
1/3	Gear	1725	0.053565	0.4	1233	3
1/3	Gear	1725	0.080348	0.6	841	3
1/3	Gear	1725	0.093739	0.7	638	3
1/3	Gear	1725	0.120522	0.9	508	3
1/3	Gear	1725	0.147304	1.1	421	3
1/3	Gear	3450	0.093739	0.7	624	3
1/3	Gear	3450	0.147304	1.1	421	3

Motor (hp)	Pump Type	Speed (rpm)	Displacement, in³/rev	Flow rate (gpm)	Pressure, continuous (psi)	Phase
½	Gear	1725	0.053565	0.4	1857	3
½	Gear	1725	0.080348	0.6	1262	3
½	Gear	1725	0.093739	0.7	957	3
½	Gear	1725	0.120522	0.9	769	3
½	Gear	1725	0.147304	1.1	638	3
½	Gear	1725	0.187478	1.4	479	3
½	Gear	1725	0.214261	1.6	435	3
½	Gear	1725	0.254435	1.9	363	3
½	Gear	3450	0.093739	0.7	928	3
½	Gear	3450	0.147304	1.1	638	3
½	Gear	3450	0.187478	1.4	479	3
½	Gear	3450	0.241043	1.8	377	3

Pump Data

Motor (hp)	Pump Type	Speed (rpm)	Displacement, in³/rev	Flow rate (gpm)	Pressure, continuous (psi)	Phase
¾	Gear	1725	0.053565	0.4	2785	3
¾	Gear	1725	0.080348	0.6	1900	3
¾	Gear	1725	0.093739	0.7	1436	3
¾	Gear	1725	0.120522	0.9	1146	3
¾	Gear	1725	0.147304	1.1	957	3
¾	Gear	1725	0.187478	1.4	725	3
¾	Gear	1725	0.214261	1.6	667	3
¾	Gear	1725	0.254435	1.9	551	3
¾	Gear	1725	0.281217	2.1	493	3
¾	Gear	1725	0.334783	2.5	406	3
¾	Gear	1725	0.361565	2.7	392	3
¾	Gear	3450	0.093739	0.7	1392	3
¾	Gear	3450	0.147304	1.1	943	3
¾	Gear	3450	0.187478	1.4	725	3
¾	Gear	3450	0.241043	1.8	566	3
¾	Gear	3450	0.294609	2.2	479	3
¾	Gear	3450	0.388348	2.9	363	3
¾	Gear	3450	0.41513	3.1	334	3

Motor (hp)	Pump Type	Speed (rpm)	Displacement, in³/rev	Flow rate (gpm)	Pressure, continuous (psi)	Phase
1	Gear	1725	0.080348	0.6	2524	3
1	Gear	1725	0.093739	0.7	1915	3
1	Gear	1725	0.120522	0.9	1523	3
1	Gear	1725	0.147304	1.1	1276	3
1	Gear	1725	0.187478	1.4	957	3
1	Gear	1725	0.214261	1.6	885	3
1	Gear	1725	0.254435	1.9	725	3
1	Gear	1725	0.281217	2.1	667	3
1	Gear	1725	0.334783	2.5	551	3
1	Gear	1725	0.361565	2.7	508	3
1	Gear	1725	0.455304	3.4	406	3
1	Gear	3450	0.093739	0.7	1857	3
1	Gear	3450	0.147304	1.1	1262	3
1	Gear	3450	0.187478	1.4	957	3
1	Gear	3450	0.241043	1.8	769	3
1	Gear	3450	0.294609	2.2	638	3
1	Gear	3450	0.388348	2.9	479	3
1	Gear	3450	0.41513	3.1	435	3
1	Gear	3450	0.50887	3.8	363	3

Pump Data

Motor (hp)	Pump Type	Speed (rpm)	Displacement, in³/rev	Flow rate (gpm)	Pressure, continuous (psi)	Phase
1½	Gear	1725	0.093739	0.7	2886	3
1½	Gear	1725	0.120522	0.9	2292	3
1½	Gear	1725	0.147304	1.1	1900	3
1½	Gear	1725	0.187478	1.4	1436	3
1½	Gear	1725	0.214261	1.6	1320	3
1½	Gear	1725	0.254435	1.9	1088	3
1½	Gear	1725	0.281217	2.1	1001	3
1½	Gear	1725	0.334783	2.5	827	3
1½	Gear	1725	0.361565	2.7	769	3
1½	Gear	1725	0.455304	3.4	609	3
1½	Gear	3450	0.093739	0.7	2785	3
1½	Gear	3450	0.147304	1.1	1900	3
1½	Gear	3450	0.187478	1.4	1436	3
1½	Gear	3450	0.241043	1.8	1146	3
1½	Gear	3450	0.294609	2.2	957	3
1½	Gear	3450	0.388348	2.9	725	3
1½	Gear	3450	0.41513	3.1	667	3
1½	Gear	3450	0.50887	3.8	551	3
1½	Gear	3450	0.562435	4.2	493	3

Motor (hp)	Pump Type	Speed (rpm)	Displacement, in³/rev	Flow rate (gpm)	Pressure, continuous (psi)	Phase
2	Gear	1725	0.120522	0.9	2901	3
2	Gear	1725	0.147304	1.1	2538	3
2	Gear	1725	0.187478	1.4	1914	3
2	Gear	1725	0.214261	1.6	1769	3
2	Gear	1725	0.254435	1.9	1450	3
2	Gear	1725	0.281217	2.1	1334	3
2	Gear	1725	0.334783	2.5	1102	3
2	Gear	1725	0.361565	2.7	1030	3
2	Gear	1725	0.455304	3.4	812	3
2	Gear	3450	0.147304	1.1	2524	3
2	Gear	3450	0.187478	1.4	1914	3
2	Gear	3450	0.241043	1.8	1523	3
2	Gear	3450	0.294609	2.2	1276	3
2	Gear	3450	0.388348	2.9	957	3
2	Gear	3450	0.41513	3.1	885	3
2	Gear	3450	0.50887	3.8	725	3
2	Gear	3450	0.562435	4.2	667	3

Pump Data

Motor (hp)	Pump Type	Speed (rpm)	Displacement, in³/rev	Flow rate (gpm)	Pressure, continuous (psi)	Phase
3	Gear	1725	**0.187478**	**1.4**	**2886**	3
3	Gear	1725	**0.214261**	**1.6**	**2654**	3
3	Gear	1725	**0.254435**	**1.9**	**2176**	3
3	Gear	1725	**0.281217**	**2.1**	2002	3
3	Gear	1725	0.334783	2.5	1653	3
3	Gear	1725	0.361565	2.7	1537	3
3	Gear	1725	0.455304	3.4	1218	3
3	Gear	3450	0.241043	1.8	2292	3
3	Gear	3450	0.294609	2.2	1900	3
3	Gear	3450	0.388348	2.9	1436	3
3	Gear	3450	0.41513	3.1	1320	3
3	Gear	3450	0.50887	3.8	1088	3
3	Gear	3450	0.562435	4.2	1001	3

Motor (hp)	Pump Type	Speed (rpm)	Displacement, in³/rev	Flow rate (gpm)	Pressure, continuous (psi)	Phase
5	Gear	1725	**0.281217**	**2.1**	2901	3
5	Gear	1725	0.334783	2.5	2741	3
5	Gear	1725	0.361565	2.7	2567	3
5	Gear	1725	0.455304	3.4	2016	3
5	Gear	3450	0.294609	2.2	2901	3
5	Gear	3450	0.388348	2.9	2393	3
5	Gear	3450	0.41513	3.1	2205	3
5	Gear	3450	0.50887	3.8	1813	3
5	Gear	3450	0.562435	4.2	1668	3

Pump Data

Motor hp	Pump type	Speed (rpm)	Displacement (in³/rev)	Pressure (Cont) (psi)
7.4	Ext Gear	1455	0.488	3480
7.4	Ext Gear	1455	0.610	2755
7.4	Ext Gear	1455	0.763	2248
7.4	Ext Gear	1455	0.976	1740
7.4	Ext Gear	1455	1.04	1595
7.4	Ext Gear	1455	1.22	1378
7.4	Ext Gear	1455	1.53	1088
7.4	Ext Gear	1455	1.65	1015
7.4	Ext Gear	1455	2.07	800

Motor hp	Pump type	Speed (rpm)	Displacement (cc/rev)	Pressure (Cont) (psi)
10	Ext Gear	1455	0.610	3625
10	Ext Gear	1455	0.763	3045
10	Ext Gear	1455	0.976	2393
10	Ext Gear	1455	1.04	2250
10	Ext Gear	1455	1.22	1885
10	Ext Gear	1455	1.53	1522
10	Ext Gear	1455	1.65	1377
10	Ext Gear	1455	2.07	1087

Pump Data

Motor hp	Pump type	Speed (rpm)	Displacement (in³/rev)	Pressure (Cont) (psi)
15.5	Ext. Gear	4000	0.32	4000
19.2	Ext. Gear	4000	0.4	4000
24.4	Ext. Gear	4000	0.5	4000
27.4	Ext. Gear	3600	0.6	4000
31.4	Ext. Gear	3300	0.8	4000
35.5	Ext. Gear	3000	1	4000
36.9	Ext. Gear	2800	1.2	3625
38.6	Ext. Gear	2600	1.5	3400
34.3	Ext. Gear	2300	1.75	3000
31.4	Ext. Gear	2100	2	2500

Pump Data – External Gear Pumps (with 1-phase Motor)

Motor (hp)	Pump Type	Speed (rpm)	Displacement, in³/rev	Flow rate (gpm)	Pressure, continuous (psi)	Ph
¼	Gear	1725	0.053565	0.4	928	1
¼	Gear	1725	0.080348	0.6	638	1
¼	Gear	1725	0.093739	0.7	479	1
¼	Gear	1725	0.120522	0.9	377	1
¼	Gear	1725	0.147304	1.1	319	1

Motor (hp)	Pump Type	Speed (rpm)	Displacement, in³/rev	Flow rate (gpm)	Pressure, continuous (psi)	Ph
1/3	Gear	1725	0.053565	0.4	1233	1
1/3	Gear	1725	0.080348	0.6	841	1
1/3	Gear	1725	0.093739	0.7	638	1
1/3	Gear	1725	0.120522	0.9	508	1
1/3	Gear	1725	0.147304	1.1	421	1
1/3	Gear	1725	0.187478	1.4	319	1
1/3	Gear	1725	0.214261	1.6	290	1
1/3	Gear	1725	0.254435	1.9	247	1
1/3	Gear	3450	0.093739	0.7	624	1
1/3	Gear	3450	0.147304	1.1	421	1
1/3	Gear	3450	0.187478	1.4	319	1
1/3	Gear	3450	0.241043	1.8	261	1

Motor (hp)	Pump Type	Speed (rpm)	Displacement, in³/rev	Flow rate (gpm)	Pressure, continuous (psi)	Ph
½	Gear	1725	0.053565	0.4	1857	1
½	Gear	1725	0.080348	0.6	1262	1
½	Gear	1725	0.093739	0.7	957	1
½	Gear	1725	0.120522	0.9	769	1
½	Gear	1725	0.147304	1.1	638	1
½	Gear	1725	0.187478	1.4	479	1
½	Gear	1725	0.214261	1.6	435	1
½	Gear	1725	0.254435	1.9	363	1
½	Gear	1725	0.281217	2.1	334	1
½	Gear	1725	0.334783	2.5	276	1
½	Gear	1725	0.361565	2.7	261	1
½	Gear	3450	0.093739	0.7	928	1
½	Gear	3450	0.147304	1.1	638	1
½	Gear	3450	0.187478	1.4	479	1

Pump Data

Motor (hp)	Pump Type	Speed (rpm)	Displacement, in³/rev	Flow rate (gpm)	Pressure, continuous (psi)	Ph
¾	Gear	1725	0.080348	0.6	1900	1
¾	Gear	1725	0.093739	0.7	1436	1
¾	Gear	1725	0.120522	0.9	1146	1
¾	Gear	1725	0.147304	1.1	957	1
¾	Gear	1725	0.187478	1.4	725	1
¾	Gear	1725	0.214261	1.6	667	1
¾	Gear	1725	0.254435	1.9	551	1
¾	Gear	1725	0.281217	2.1	493	1
¾	Gear	1725	0.334783	2.5	406	1
¾	Gear	1725	0.361565	2.7	392	1
¾	Gear	1725	0.455304	3.4	305	
¾	Gear	3450	0.093739	0.7	1392	1
¾	Gear	3450	0.147304	1.1	943	1
¾	Gear	3450	0.187478	1.4	725	1
¾	Gear	3450	0.241043	1.8	566	1
¾	Gear	3450	0.294609	2.2	479	1
¾	Gear	3450	0.388348	2.9	363	1
¾	Gear	3450	0.41513	3.1	334	1
¾	Gear	3450	0.50887	3.8	276	1

Motor (hp)	Pump Type	Speed (rpm)	Displacement, in³/rev	Flow rate (gpm)	Pressure, continuous (psi)	Ph
1	Gear	1725	0.093739	0.7	1915	1
1	Gear	1725	0.120522	0.9	1523	1
1	Gear	1725	0.147304	1.1	1276	1
1	Gear	1725	0.187478	1.4	957	1
1	Gear	1725	0.214261	1.6	885	1
1	Gear	1725	0.254435	1.9	725	1
1	Gear	1725	0.281217	2.1	667	1
1	Gear	1725	0.334783	2.5	551	1
1	Gear	1725	0.361565	2.7	508	1
1	Gear	1725	0.455304	3.4	406	1
1	Gear	3450	0.093739	0.7	1857	1
1	Gear	3450	0.147304	1.1	1262	1
1	Gear	3450	0.187478	1.4	957	1
1	Gear	3450	0.241043	1.8	769	1
1	Gear	3450	0.294609	2.2	638	1
1	Gear	3450	0.388348	2.9	479	1
1	Gear	3450	0.41513	3.1	435	1
1	Gear	3450	0.50887	3.8	363	1

Motor (hp)	Pump Type	Speed (rpm)	Displacement, in³/rev	Flow rate (gpm)	Pressure, continuous (psi)	Ph
1½	Gear	1725	0.120522	0.9	2292	1
1½	Gear	1725	0.147304	1.1	1900	1
1½	Gear	1725	0.187478	1.4	1436	1
1½	Gear	1725	0.214261	1.6	1320	1
1½	Gear	1725	0.254435	1.9	1088	1
1½	Gear	1725	0.281217	2.1	1001	1
1½	Gear	1725	0.334783	2.5	827	1
1½	Gear	1725	0.361565	2.7	769	1
1½	Gear	1725	0.455304	3.4	609	1
1½	Gear	3450	0.093739	0.7	2785	1
1½	Gear	3450	0.147304	1.1	1900	1
1½	Gear	3450	0.187478	1.4	1436	1
1½	Gear	3450	0.241043	1.8	1146	1
1½	Gear	3450	0.294609	2.2	957	1
1½	Gear	3450	0.388348	2.9	725	1
1½	Gear	3450	0.41513	3.1	667	1
1½	Gear	3450	0.50887	3.8	551	1
1½	Gear	3450	0.562435	4.2	493	1

Motor (hp)	Pump Type	Speed	Displacement, in³/rev	Flow rate (gpm)	Pressure, continuous (psi)	Ph
2	Gear	1725	0.120522	0.9	2901	1
2	Gear	1725	0.187478	1.4	1914	1
2	Gear	1725	0.214261	1.6	1769	1
2	Gear	1725	0.254435	1.9	1450	1
2	Gear	1725	0.281217	2.1	1334	1
2	Gear	1725	0.334783	2.5	1102	1
2	Gear	1725	0.361565	2.7	1030	1
2	Gear	1725	0.455304	3.4	812	1
2	Gear	3450	0.147304	1.1	2524	1
2	Gear	3450	0.187478	1.4	1914	1
2	Gear	3450	0.241043	1.8	1523	1
2	Gear	3450	0.294609	2.2	1276	1
2	Gear	3450	0.388348	2.9	957	1
2	Gear	3450	0.41513	3.1	885	1
2	Gear	3450	0.50887	3.8	725	1
2	Gear	3450	0.562435	4.2	667	1

Pump Data

Motor (hp)	Pump Type	Speed (rpm)	Displacement, in³/rev	Flow rate (gpm)	Pressure, continuous (psi)	Ph
3	Gear	1725	0.254435	1.9	2176	1
3	Gear	1725	0.281217	2.1	2002	1
3	Gear	1725	0.334783	2.5	1653	1
3	Gear	1725	0.361565	2.7	1537	1
3	Gear	1725	0.455304	3.4	1218	1
3	Gear	3450	0.187478	1.4	2886	1
3	Gear	3450	0.241043	1.8	2901	1
3	Gear	3450	0.294609	2.2	1900	1
3	Gear	3450	0.388348	2.9	1436	1
3	Gear	3450	0.41513	3.1	1320	1
3	Gear	3450	0.50887	3.8	1088	1
3	Gear	3450	0.562435	4.2	1001	1

Pump Data – Vane Pumps

Motor (hp)*	Pump Type	Max. Speed (rpm)	Displacement, in³/rev	Flow rate (gpm)*	Max. Pressure, continuous (psi)
7	Vane	4800	0.2	3.6	2500
13.6	Vane	4500	0.4	7.3	2500
17.8	Vane	4000	0.6	9.4	2500
20.4	Vane	3400	0.8	10.9	2500
22.8	Vane	3200	1	12.8	2500
24	Vane	2800	1.39	16	2000
24.5	Vane	3000	1.19	14.6	2000
29	Vane	3400	1.19	16.1	2500
29.5	Vane	3000	1.39	16.7	2500
32.5	Vane	2800	1.62	17.7	2500
35.5	Vane	2800	1.81	19.8	2500
36	Vane	2400	2.38	23	2000
37.5	Vane	2500	2.22	22.9	2500
39	Vane	2400	2.59	25.9	2000

*@ Max. speed and pressure

Pump Data – Piston Pumps

Motor (hp)*	Pump Type	Max. Speed (rpm)	Displacement, in³/rev	Flow rate (gpm)*	Max. Pressure, continuous (psi)
16.3	Piston	3000	0.66	4.2	5000
17.7	Piston	3000	0.86	5.9	4000
20.2	Piston	3000	1.35	9.5	3000
35.0	Piston	2400	2.83	20.6	2500
35.5	Piston	3000	2.06	14.7	3500
36.5	Piston	3000	1.55	10.9	5000
64	Piston	1800	7.94	58.2	1500
73	Piston	2700	2.93	21.1	5000
74	Piston	2400	6.00	43.3	2500
80	Piston	2400	4.67	33.7	3500
89	Piston	2700	4.6	33.3	3750
95	Piston	2400	3.88	27.4	5000
100	Piston	2700	3.98	28.8	5000
150	Piston	2400	6.0	42.4	5000
150	Piston	2400	7.94	57.6	3750
215	Piston	2400	9.16	63.0	5000

*@ 1800 rpm, rated constant pressure

Appendix 2

The data is extracted from the actual catalogue listing of some prominent manufacturers.

Performance Data of Gerolor Motors for continuous duty

Table A2.1 | Values of Torque in lb.in (Speed in rpm)

$V_D = 0.50$ in^3/rev							
ΔP(psi) Q(gpm)	200	400	600	800	1000	1500	2030
1	11 (456)	25 (444)	40 (429)	55 (412)	69 (394)	102 (332)	132 (239)
2	9 (897)	24 (886)	38 (867)	53 (847)	68 (823)	105 (749)	141 (647)
3	6 (1349)	20 (1331)	35 (1309)	51 (1285)	65 (1261)	102 (1176)	139 (1060)
4.25		16 (1902)	30 (1873)	44 (1846)	60 (1817)	97 (1721)	135 (1585)
4.5		16 (1992)	29 (1964)	43 (1929)	59 (1900)	96 (1808)	134 (1673)

Table A2.2 | Values of Torque in lb.in (Speed in rpm)

$V_D = 0.79$ in^3/rev							
ΔP(psi) Q(gpm)	200	400	600	800	1000	1500	2030
1	19 (290)	43 (285)	65 (277)	88 (268)	109 (260)	164 (230)	217 (189)
2	16 (573)	39 (566)	63 (555)	86 (544)	109 (534)	165 (490)	225 (437)
3	11 (859)	35 (849)	58 (838)	82 (825)	105 (810)	163 (763)	223 (701)
4	6 (1153)	30 (1140)	53 (1129)	76 (1117)	99 (1101)	157 (1044)	217 (975)
5.5		19 (1575)	42 (1556)	65 (1539)	89 (1521)	148 (1457)	209 (1387)

Performance Data of Gerolor Motors for continuous duty

Table A2.3 | Values of Torque in lb.in (Speed in rpm)

ΔP(psi) Q(gpm)	$V_D = 1.21$ in³/rev						
	200	400	600	800	1000	1500	2030
1	32 (189)	67 (187)	102 (185)	136 (182)	170 (179)	253 (169)	325 (138)
2	30 (379)	65 (375)	101 (370)	136 (366)	172 (361)	257 (347)	333 (309)
3	21 (569)	57 (565)	93 (560)	128 (556)	163 (551)	248 (523)	330 (484)
4	12 (761)	47 (758)	83 (751)	119 (746)	154 (741)	239 (707)	320 (656)
5.5		31 (1043)	67 (1035)	101 (1028)	137 (1021)	218 (990)	299 (934)

Table A2.4 | Values of Torque in lb.in (Speed in rpm)

ΔP(psi) Q(gpm)	$V_D = 1.93$ in³/rev						
	200	400	600	800	1000	1500	1750
1	51 (118)	106 (116)	160 (113)	213 (111)	265 (107)	383 (81)	439 (70)
2	46 (236)	103 (234)	159 (230)	214 (225)	269 (221)	387 (175)	446 (165)
3	36 (355)	94 (352)	149 (347)	205 (342)	259 (336)	377 (287)	440 (273)
4	24 (474)	79 (472)	135 (466)	190 (460)	246 (452)	362 (393)	425 (373)
5.5		55 (650)	111 (645)	167 (636)	221 (629)	334 (575)	400 (550)

Table A2.5 | Values of Torque in lb.in (Speed in rpm)

ΔP(psi) Q(gpm)	$V_D = 3.0$ in³/rev						
	200	400	600	800	1000	1200	1400
1	82 (75)	167 (72)					
2	70 (149)	156 (147)	243 (144)	327 (142)			
3	53 (221)	140 (220)	227 (217)	311 (213)	396 (209)	484 (201)	549 (191)
4	30 (296)	120 (292)	204 (286)	292 (282)	374 (273)	460 (265)	541 (259)
5.5		81 (393)	170 (389)	254 (383)	339 (377)	422 (369)	506 (358)

Performance Data of Gerolor Motors for continuous duty

Table A2.6 | Values of Torque in lb.in (Speed in rpm)

ΔP(psi) Q(gpm)	$V_D = 2.2$ in³/rev						
	200	400	600	800	1000	1400	1800
2	49 (204)	103 (201)	162 (198)	189 (194)	270 (189)	379 (177)	489 (162)
4	47 (408)	106 (407)	160 (402)	191 (399)	274 (394)	384 (381)	495 (365)
6	44 (613)	102 (612)	158 (609)	188 (604)	272 (599)	383 (586)	496 (565)
8	40 (817)	97 (817)	153 (814)	184 (807)	270 (799)	383 (785)	497 (762)
10	36 (1021)	90 (1021)	148 (1015)	180 (1008)	265 (1001)	380 (981)	495 (959)

Table A2.7 | Values of Torque in lb.in (Speed in rpm)

ΔP(psi) Q(gpm)	$V_D = 2.8$ in³/rev						
	200	400	600	800	1000	1400	1800
2	64 (161)	136 (158)	212 (156)	284 (153)	355 (148)	497 (139)	641 (127)
4	61 (323)	139 (320)	209 (316)	286 (314)	359 (310)	503 (300)	649 (287)
6	58 (486)	134 (481)	207 (479)	282 (475)	356 (471)	502 (461)	650 (444)
8	52 (648)	128 (643)	200 (640)	276 (635)	354 (628)	502 (617)	651 (599)
10	47 (808)	118 (803)	194 (798)	269 (793)	347 (787)	498 (771)	649 (753)
12	36 (969)	109 (964)	188 (960)	260 (952)	340 (946)	492 (931)	643 (914)

Performance Data of Gerolor Motors for continuous duty

Table A2.8 | Values of Torque in lb.in (Speed in rpm)

ΔP(psi) Q(gpm)	200	400	600	800	1000	1400	1800
	V_D = 3.6 in³/rev						
2	79 (127)	169 (125)	260 (123)	305 (121)	437 (117)	616 (109)	796 (96)
4	76 (254)	168 (254)	257 (251)	307 (249)	441 (246)	620 (236)	800 (224)
6	73 (381)	161 (381)	252 (380)	303 (377)	439 (373)	618 (364)	802 (349)
8	64 (508)	151 (508)	243 (508)	294 (504)	428 (500)	609 (491)	794 (476)
10	57 (635)	141 (635)	234 (634)	283 (630)	419 (626)	602 (614)	786 (601)
12	45 (762)	131 (762)	227 (762)	274 (757)	409 (753)	593 (741)	778 (728)
14	33 (889)	118 (889)	213 (887)	266 (882)	396 (877)	583 (866)	770 (851)
15	29 (953)	111 (953)	205 (951)	260 (945)	389 (940)	576 (929)	765 (913)

Table A2.9 | Values of Torque in lb.in (Speed in rpm)

ΔP(psi) Q(gpm)	200	400	600	800	1000	1400	1800
	V_D = 4.5 in³/rev						
2	103 (101)	220 (99)	339 (98)	454 (96)	569 (93)	801 (86)	1036 (76)
4	99 (203)	219 (201)	335 (199)	457 (197)	574 (194)	808 (187)	1042 (177)
6	94 (305)	210 (303)	328 (301)	451 (298)	571 (296)	805 (288)	1044 (276)
8	86 (406)	196 (404)	319 (402)	438 (399)	558 (396)	793 (388)	1033 (377)
10	74 (507)	183 (505)	310 (502)	422 (499)	545 (496)	784 (486)	1024 (476)
12	58 (608)	171 (606)	295 (603)	408 (600)	533 (596)	773 (587)	1013 (576)
14	43 (709)	154 (706)	277 (702)	396 (698)	515 (694)	760 (686)	1002 (674)
15	36 (760)	145 (757)	268 (753)	387 (749)	506 (744)	750 (735)	996 (723)

Performance Data of Gerolor Motors for continuous duty

Table A2.10 | Values of Torque in lb.in (Speed in rpm)

ΔP(psi) \ Q(gpm)	200	400	600	800	1000	1400	1800
				$V_D = 5.9$ in³/rev			
2	134 (78)	292 (76)	442 (75)	593 (73)	746 (71)	1054 (65)	1365 (55)
4	131 (156)	281 (155)	436 (153)	596 (151)	750 (149)	1059 (143)	1367 (134)
6	126 (234)	269 (233)	425 (231)	588 (230)	747 (228)	1054 (221)	1368 (210)
8	110 (312)	246 (311)	408 (310)	566 (308)	718 (305)	1023 (300)	1339 (291)
10	96 (390)	231 (389)	392 (387)	539 (385)	699 (383)	1005 (376)	1318 (368)
12	77 (468)	218 (467)	378 (465)	522 (463)	681 (460)	990 (453)	1301 (445)
14	60 (546)	197 (544)	358 (542)	513 (539)	662 (537)	973 (531)	1293 (521)
15	52 (585)	189 (583)	346 (581)	495 (578)	651 (575)	963 (589)	1286 (559)

Table A2.11 | Values of Torque in lb.in (Speed in rpm)

ΔP(psi) \ Q(gpm)	200	400	600	800	1000	1400	1800
				$V_D = 7.3$ in³/rev			
2	162 (62)	357 (61)	544 (61)	736 (59)	927 (58)	1305 (53)	1687 (45)
4	160 (125)	348 (124)	539 (123)	736 (121)	930 (120)	1316 (116)	1698 (110)
6	155 (188)	338 (187)	530 (186)	729 (185)	923 (183)	1310 (178)	1699 (170)
8	139 (250)	319 (250)	515 (249)	710 (247)	901 (245)	1283 (241)	1673 (233)
10	121 (313)	303 (312)	497 (311)	686 (309)	883 (308)	1267 (302)	1655 (296)
12	102 (375)	288 (374)	480 (373)	664 (371)	862 (370)	1246 (365)	1640 (358)
14	78 (438)	263 (437)	458 (435)	652 (433)	841 (431)	1228 (427)	1616 (419)
15	67 (469)	253 (468)	446 (466)	632 (464)	828 (462)	1214 (458)	1608 (450)

Performance Data of Gerolor Motors for continuous duty

Table A2.12 | Values of Torque in lb.in (Speed in rpm)

ΔP(psi) Q(gpm)	$V_D = 8.9$ in³/rev						
	200	400	600	800	1000	1400	1700
2	198 (51)	435 (50)	664 (50)	897 (49)	1130 (47)	1591 (43)	1942 (39)
4	196 (103)	424 (102)	657 (101)	898 (99)	1133 (99)	1604 (95)	1954 (92)
6	189 (154)	412 (153)	646 (152)	889 (151)	1125 (150)	1598 (146)	1951 (141)
8	169 (205)	389 (205)	628 (204)	866 (203)	1098 (201)	1564 (197)	1919 (193)
10	148 (257)	369 (256)	605 (255)	836 (253)	1076 (252)	1544 (248)	1899 (244)
12	125 (308)	351 (307)	586 (306)	810 (305)	1051 (303)	1519 (299)	1878 (295)
14	95 (359)	321 (358)	558 (357)	795 (355)	1026 (354)	1497 (350)	1851 (346)
15	82 (385)	308 (384)	544 (383)	771 (381)	1010 (379)	1480 (375)	1840 (371)

Table A2.13 | Values of Torque in lb.in (Speed in rpm)

ΔP(psi) Q(gpm)	$V_D = 9.7$ in³/rev						
	200	400	600	800	1000	1400	1650
2	209 (47)	465 (46)	715 (46)	973 (45)	1228 (44)	1724 (40)	2046 (37)
4	210 (94)	460 (94)	710 (93)	971 (91)	1229 (91)	1745 (89)	2059 (87)
6	205 (141)	454 (141)	704 (140)	965 (139)	1216 (138)	1738 (134)	2055 (132)
8	186 (188)	440 (188)	693 (187)	951 (186)	1205 (185)	1716 (181)	2038 (178)
10	164 (235)	422 (234)	671 (234)	930 (232)	1189 (232)	1702 (228)	2032 (225)
12	144 (282)	404 (281)	652 (281)	900 (279)	1163 (279)	1674 (275)	2004 (272)
14	109 (330)	374 (329)	623 (328)	883 (327)	1140 (325)	1653 (322)	1963 (319)
15	92 (353)	359 (352)	612 (351)	861 (350)	1123 (348)	1633 (345)	1950 (342)

Performance Data of Gerolor Motors for continuous duty

Table A2.14 | Values of Torque in lb.in (Speed in rpm)

$\Delta P(psi)$ Q(gpm)	$V_D = 11.3$ in³/rev						
	200	400	600	800	1000	1400	1700
2	257 (40)	554 (40)	847 (39)	1150 (38)	1447 (37)	2035 (33)	2320 (29)
4	254 (81)	546 (81)	845 (80)	1145 (79)	1448 (78)	2049 (76)	2343 (74)
6	246 (121)	540 (121)	834 (120)	1137 (120)	1434 (119)	2036 (115)	2337 (112)
8	224 (162)	520 (162)	820 (161)	1117 (160)	1414 (159)	2014 (155)	2315 (152)
10	202 (202)	499 (202)	793 (201)	1095 (201)	1394 (200)	1997 (196)	2299 (193)
12	176 (243)	475 (242)	767 (242)	1063 (241)	1368 (240)	1969 (236)	2268 (234)
14	140 (283)	443 (283)	735 (282)	1035 (281)	1340 (280)	1936 (277)	2227 (274)
15	120 (304)	425 (303)	719 (302)	1014 (301)	1320 (300)	1914 (297)	2205 (294)

Table A2.15 | Values of Torque in lb.in (Speed in rpm)

$\Delta P(psi)$ Q(gpm)	$V_D = 14.1$ in³/rev						
	200	400	600	800	1000	1400	1450
2	338 (32)	707 (32)	1074 (31)	1456 (30)	1827 (30)	2572 (26)	2657 (25)
4	328 (65)	695 (65)	1076 (64)	1447 (63)	1827 (62)	2577 (60)	2669 (60)
6	317 (97)	687 (97)	1057 (97)	1434 (96)	1811 (95)	2555 (92)	2650 (91)
8	289 (130)	659 (130)	1038 (130)	1406 (129)	1777 (128)	2531 (124)	2625 (124)
10	265 (162)	631 (162)	1004 (162)	1381 (162)	1751 (160)	2510 (156)	2602 (156)
12	230 (195)	599 (195)	968 (194)	1345 (194)	1722 (193)	2480 (189)	2571 (189)
14	191 (227)	563 (227)	927 (227)	1299 (226)	1686 (226)	2428 (222)	2519 (221)
15	167 (243)	538 (243)	904 (243)	1279 (242)	1661 (242)	2404 (238)	2493 (238)

Performance Data of Gerolor Motors for continuous duty

Table A2.16 | Values of Torque in lb.in (Speed in rpm)

$V_D = 17.9$ in³/rev							
ΔP(psi) Q(gpm)	200	400	600	800	1000	1200	1350
2	427 (26)	893 (25)	1361 (25)	1829 (24)	2293 (22)	2672 (16)	2977 (13)
4	419 (51)	886 (51)	1362 (51)	1833 (50)	2305 (49)	2771 (47)	3110 (44)
6	402 (77)	872 (77)	1342 (76)	1819 (76)	2291 (74)	2757 (71)	3098 (68)
8	367 (102)	838 (102)	1316 (102)	1785 (101)	2252 (100)	2723 (98)	3070 (95)
10	332 (128)	803 (128)	1276 (128)	1749 (127)	2215 (126)	2684 (123)	3034 (120)
12	289 (153)	760 (153)	1230 (153)	1706 (153)	2177 (151)	2634 (149)	2989 (146)
14	241 (179)	712 (179)	1176 (179)	1650 (179)	2126 (177)	2592 (175)	2935 (172)
15	211 (192)	683 (192)	1149 (192)	1623 (191)	2096 (190)	2558 (188)	2905 (185)

Table A2.17 | Values of Torque in lb.in (Speed in rpm)

$V_D = 22.6$ in³/rev							
ΔP(psi) Q(gpm)	200	400	600	800	1000	1200	1250
2	537 (20)	1121 (20)	1715 (20)	2285 (19)	2862 (16)		
4	532 (40)	1123 (40)	1715 (40)	2308 (39)	2893 (38)	3467 (36)	3604 (35)
6	508 (61)	1100 (61)	1693 (61)	2294 (60)	2884 (58)	3458 (55)	3598 (53)
8	463 (81)	1060 (81)	1661 (81)	2255 (80)	2840 (79)	3414 (76)	3557 (74)
10	414 (101)	1017 (101)	1613 (101)	2203 (101)	2788 (99)	3363 (96)	3506 (94)
12	363 (121)	960 (121)	1553 (121)	2152 (121)	2737 (119)	3305 (116)	3446 (115)
14	303 (142)	897 (142)	1484 (142)	2086 (142)	2667 (140)	3246 (137)	3386 (136)
15	266 (152)	862 (152)	1452 (152)	2050 (152)	2630 (150)	3206 (148)	3347 (147)

Performance Data of Gerolor Motors for continuous duty

Table A2.18 | Values of Torque in lb.in (Speed in rpm)

$\Delta P(psi)$ $Q(gpm)$	$V_D = 45.1\ in^3/rev$		
	200	400	600
2	1080 (10)	2250 (10)	3440 (10)
4	1070 (20)	2250 (20)	3440 (19)
6	1020 (30)	2200 (30)	3390 (29)
8	945 (40)	2135 (40)	3330 (39)
10	840 (50)	2050 (50)	3250 (48)
12	740 (60)	1945 (59)	3130 (58)
14	630 (69)	1820 (68)	3005 (68)
15	540 (74)	1735 (74)	2905 (73)

Performance Data of Bent-axis Axial Piston motors

Table A2.19 |

Displacement, in^3/rev	Maximum pressure, psi (Continuous)	Maximum speed, rpm	Motor continuous input flow, gpm
0.299	5000	10800	10.83
0.598	5000	9900	17.70
0.873	5000	9000	22.99
1.159	5000	8100	27.21
1.830	5000	5600	44.39
2.440	5000	5000	52.84
3.649	5000	4300	67.90
4.906	5000	4000	85.07
6.719	5000	3600	104.62
9.154	5000	2600	103.04
14.768	5000	2400	153.24

Performance Data of Gear Motors

Table A2.20

Output Power (hp)	Displacement (in³/rev)	Speed, max (rpm)	Flow Rate, max (gpm)	Pressure drop, max (psi)	Inlet pressure, max (psi)	Torque, max (lb.in)
2.4	0.5	1950	4.2	1500	2030	106
2.48	3.05	400	5.5	1015	2030	398
2.5	2.44	500	5.5	1200	2030	375
3.3	0.79	1550	5.5	1500	2030	150
3.3	1.22	1000	5.5	1500	2030	230
3.3	1.93	630	5.5	1500	2030	375
4.4	38.05	95	15.9	55	2030	3895
4.7	30.2	120	15.9	60	2030	3452
6.0	1.52	1600	10.5	1450	2540	290
6.2	24.16	150	15.9	70	2540	3190
7.8	1.95	1560	13.2	1450	2540	380
7.9	19.3	190	15.9	90	2540	3360
10	15.1	242	15.9	110	2540	3360
11.5	2.44	1500	15.9	1750	2540	550
13.5	3.02	1210	15.9	2030	2540	835
13.5	9.66	378	15.9	140	2540	2770
13.5	12.1	303	15.9	140	2540	3240
13.7	4.83	755	15.9	2030	2540	1340
13.7	7.55	486	15.9	2030	2540	2100
14.1	6.04	605	15.9	2030	2540	1710

Appendix 3

A3.1| Bore Size and Piston-rod Size, Hydraulic Cylinders

Table A3.1

Bore (inch)	Rod Diameter (inch)
1.0	0.500
	0.625
1.5	0.625
	1.000
2.0	0.625
	1.375
	1.000
2.5	1.000
	1.750
	1.375
	0.625
3.25	1.000
	2.000
	1.375
	1.750
4.0	1.375
	2.500
	1.750
	2.000
	1.000
5.0	1.750
	3.500
	2.000
	2.500
	3.000
	1.000
6.0	1.750
	4.000
8.0	2.000
	5.500

A3.2 | Theoretical Cylinder Forces in the English Units
Force (lb) = Pressure (psi) x Piston Area (in²)

Table A3.2 | Theoretical Cylinder Forces in the English units

Bore size (in)	Rod dia. (in)	Force (lb)	System pressure (psi)			
			1000	2000	3000	5000
1½	¾	Thrust	1,770	3,530	5,300	8,830
		Pull	1,320	2,650	3,970	6,630
2	1	Thrust	3,140	6,280	9,420	15,700
		Pull	2,360	4,710	7,070	11,780
3	2	Thrust	7,070	14,130	21,200	35,340
		Pull	3,930	7,850	11,780	19,630
4	3	Thrust	12,550	25,130	37,700	62,830
		Pull	5,500	11,000	16,490	27,490
6	4	Thrust	28,270	56,550	84,820	141,370
		Pull	15,700	31,410	47,120	78,540
8	5	Thrust	50,260	100,530	150,800	251,330
		Pull	30,630	61,260	91,890	153,150
10	6	Thrust	78,540	158,080	235,620	392,700
		Pull	50,260	100,530	150,800	251,330

Example

Bore diameter	= 6 in
Rod diameter	= 4 in
Pressure, P	= 1000 psi
Piston area, A_{ext}	= $\prod D^2/4$ = 3.14 x $6^2/4$ = 28.26 in²

Effective area for pull stroke, $A_{ret} = \prod (D^2 - d^2)/4$
$$= 3.14 (6^2 - 4^2)/4 = 15.7 \text{ in}^2$$

Thrust	= P x Aext
	= 1000 x 28.26
	= 28260 lb
Pull	= P x Aret = 1000 x 15.7
	= 15700 lb

A3.3 | Different types of Piston Seals

- Lip Seal
- Spring-Loaded PTFE Seal
- Magnetic Piston, stainless steel cylinder body, single bi-directional piston seal
- Magnetic Piston, carbon steel body, single bi-directional piston seal
- Magnetic Piston, Aluminum Tube

A3.4 | Ports

- SAE Straight Thread O-Ring
- NPTF Ports (Dry Seal Pipe Thread)
- BSP Ports (Parallel Thread ISO 228)
- BSPT Ports (Taper Thread)
- Metric Thread Ports
- Metric Thread Ports per ISO 6149

A3.5 | Mounting Styles

- Tie Rods Extended Head-end
- Tie Rods Extended Cap-end
- Tie Rods Extended at Both Ends
- Head Rectangular Flange
- Head Square Flange
- Cap Rectangular Flange
- Cap Square Flange
- Side Lug
- Side Tapped
- Cap Fixed Clevis
- Head Trunnion
- Cap Trunnion
- Intermediate Fixed Trunnion
- Spherical Bearing

A3.6 | Important specifications to be considered while selecting hydraulic cylinders

Table A3.3

Bore size	Should satisfy the requirement of thrust/pull for the operating pressure
Piston-rod diameter	Should prevent piston rod buckling
Single rod / Double rod	
Cushions	Yes / No If yes, Head-end, Cap-end, or both ends?
Stop tube	Yes / No
Piston and piston-rod Seal type	Fluid and temperature compatibility?
Stroke length	
Piston-rod-end thread style	
Port size	For a given speed requirement
Port position	
Mounting style	
Piston rod and mounting accessories	To attach the cylinder to the load
Optional accessories	
Fluid medium	

A3.7 | Seal Materials and their Temperature Ranges

Table A3.4

Material	Temperature Range
Nitrile	-22°F to 212°F
H-Nitrile (Hydrogenated Nitrile)	-30°F to 302°F
Viton	-4°F to 400 °F
Silicone	-75°F to 450°F
EPDM	-65°F to 350°F
Polyurethane	-22°F to 230°F
Nylon	-40°F to 248°F
Teflon, virgin	-328°F to 500°F
Teflon, filled	-328°F to 500°F

Appendix 4

A4.1 | Properties and standards of typical steel materials

Table A4.1 | Properties and standards of typical steel materials

Pipe material	Properties	Standards
Cold-drawn seamless carbon steel	High-pressure capability, precise dimensions/shape, clean inside surface with no scale, excellent scaling surface after roll flaring	DIN EN 10305-4 E 355N (St. 52.4 NBK) E 235N (St. 37.4 NBK)
Cold-drawn seamless stainless steel	High-pressure capability, precise dimensions/shape, excellent scaling surface after roll flaring	DIN EN 10216-5 ASTM A269/A213 ASTM A312

A4.2 | Tensile Strengths and Yield Strengths of Steels

Table A4.2 | Tensile strengths and yield strengths of steels

Steel type	Tensile strength (N/mm²)	Yield strength (N/mm² min)
E235N tubes (St 37.4)	340	235
E355N tubes (St 52.4)	490	355
AISI 316L metric size tubes	485	170
TP 316L schedule size pipes	485	170
DIN 2391 St. 45	570	255
DIN 2391 St. 52	630	355

E – Steel for machine parts
235 – Minimum yield strength in N/mm²

A4.3 | Nominal Pipe Size (NPS), Pipes

Table A4.3 | Standard nominal pipe sizes and dimensions

Nominal Pipe Size		Outside Diameter	Wall thickness		
			Schedule 40	Schedule 80	Schedule 160
--	inch	inch	inch	inch	inch
⅛	0.125	0.405	0.068	0.095	--
¼	0.250	0.540	0.088	0.119	--
⅜	0.375	0.675	0.091	0.126	--
½	0.500	0.840	0.109	0.147	0.188
¾	0.750	1.050	0.113	0.154	0.219
1	1.000	1.315	0.133	0.179	0.250
1¼	1.250	1.660	0.140	0.191	0.250
1½	1.500	1.900	0.145	0.200	0.281
2	2.000	2.375	0.154	0.218	0.344
2½	2.500	2.875	0.203	0.276	0.375
3	3.000	3.500	0.216	0.300	0.438
3½	3.500	4.000	0.226	0.318	--
4	4.000	4.500	0.237	0.337	0.531
5	5.000	5.563	0.258	0.375	0.625
6	6.000	6.625	0.280	0.432	0.719
8	8.000	8.625	0.322	0.500	0.906
10	10.00	10.75	0.365	0.594	1.125
12	12.00	12.75	0.406	0.688	1.312
14	14.00	14.00	0.438	0.750	1.406
16	16.00	16.00	0.500	0.844	1.594
18	18.00	18.00	0.562	0.938	1.781

A4.4 | Diameter Nominal (DN), Pipes

Table A4.4 | Standard metric pipe sizes and dimensions

Nominal size	Outside Dia, mm	Wall thickness, mm				
		A	B	C	D	E
6	10.2	1.6				
8	13.5	1.8				
10	17.2	1.8				
15	21.3	2.0	2.8			
20	26.9	2.0	2.8			
25	33.7	2.0	3.2	4.2	6.3	6.3
32	42.4	2.3	3.5	4.2	6.3	6.3
40	48.3	2.3	3.5	4.2	6.3	6.3
50	60.3	2.3	3.8	4.2	6.3	6.3
65	76.1	2.6	4.2	4.2	6.3	7.0
80	88.9	2.9	4.2	4.2	7.1	7.6
90	101.6	2.9	4.5	4.5	7.1	8.1
100	114.3	3.2	4.5	4.5	8.0	8.6
125	139.7	3.6	4.5	4.5	8.0	9.5
150	168.3	4.0	4.5	4.5	8.8	11.0
200	219.1	4.5	5.8	5.8	8.8	12.5
450	457.0	6.3	6.3	6.3	8.8	12.5

A4.5 | Specifications of Tubing

Size and Pressure Chart for Carbon Steel hydraulic tubing (Inch Sizes)
Table A4.5

Tube OD (inch)	Wall thickness (inch)	Max. Working Pressure @ 6:1 SF (psi)	Burst Pressure (psi)
3/16	0.035	3422	20533
1/4	0.035	2567	15400
1/4	0.049	3593	21560
5/16	0.049	2875	17248
5/16	0.065	3813	22880
3/8	0.049	2396	14373
3/8	0.065	3178	19067
1/2	0.049	1797	10780
1/2	0.065	2383	14300
1/2	0.083	3043	18260
5/8	0.065	1907	11440
5/8	0.095	2787	16720
3/4	0.049	1198	7187
3/4	0.065	1589	9533
3/4	0.095	2322	13933
3/4	0.109	2664	15987
1	0.065	1192	7150
1	0.095	1742	10450
1	0.120	2200	13200
1-1/4	0.095	1393	8360
1-1/4	0.120	1760	10560

A4.6 | Specifications of Hoses

Table A4.6 | Typical parameters of Hoses in English units

Dash Number	ID Inch	Work pressure psi	Min. Burst pressure psi	Min. bend radius inches
-2	1/8	3000	16000	
-3	3/16	3000	16000	
-4	1/4	3000	16000	1.5
-5	5/16	3000	16000	
-6	3/8	3000	16000	2.5
-8	1/2	3000	16000	2.9
-10	5/8	3000	16000	3.3
-12	3/4	3000	16000	4.0
-14	7/8	3000	16000	
-16	1	3000	16000	5.0
-20	1 ¼	3000	16000	12.0
-24	1½	3000	12000	14.0
-32	2	3000	12000	
-36	2 ¼	3000	12000	
-40	2½	3000	12000	
-48	3	3000	12000	
-56	3½	3000	12000	
-64	4	3000	12000	
-72	4½	3000	12000	

Appendix 5

Viscosity Comparison, Typical Values

Table A5.1 | Viscosity Comparison Table

ISO-VG	CentiStoke (@40°C) (Tolerance: ±10%	CentiPoise	SSU
2	2	1.76	31
3	3	2.64	35
5	5	4.40	40
7	7	6.16	50
10	10	8.80	60
15	15	13.20	80
22	22	19.36	90
32	32	28.16	150
46	46	40.48	200
68	68	59.84	300
100	100	88.00	500
150	150	132.00	750
220	220	193.60	1000
320	320	281.60	1500
460	460	404.80	2000
680	680	598.40	3000
1000	1000	880.00	4000

Appendix 6

Recommended Filtration Levels of Hydraulic Fluids

Table A6.1

Recommended Filtration							
ISO Code	14/12/9	15/13/10	16/14/11	17/15/12	18/16/13	19/17/14	20/18/15
NAS Code	3	4	5	6	7	8	9
Absolute Filtration	$\beta_{5©} \geq 100$ (≥99%)		$\beta_{7©} \geq 100$		$\beta_{10©} \geq 100$	$\beta_{15©} \geq 100$	$\beta_{20©} \geq 100$
Components:							
Servo valves	●	●	●				
Proportional valves		●	●	●			
Variable disp pumps			●	●	●		
Cartridge valves				●	●	●	
Piston pumps/Motors				●	●	●	
Vane pumps/Motors					●	●	●
Gear pumps/Motors					●	●	●
FCVs/PCVs					●	●	●
Solenoid valves					●	●	●

Appendix 7

The SAE Aerospace Standard AS4059
The NAS 1638 standard is inactive for new components or systems after May 30, 2001 due to the changes in the ISO standards for the calibration of automatic particle counters (APCs). The SAE aerospace standard AS4059, for specifying particulate contamination in hydraulic fluids in different classes, was developed in 1988 as a replacement to the NAS 1638 standard. Since then this standard has undergone many revisions.

This standard AS4059 offers two classifications:

- One classification, based on the microscopic counting, applies to those currently using NAS 1638 classes and desiring to maintain the same NAS format.

- The second one, based on the automatic particle counting, applies to those using the methods of previous revisions of AS4059 and/or cumulative particle counts.

Method of Particle Counting
The introduction of automatic particle counters (APCs) during the 1960s revolutionized the measurement of the size distribution of dirt particles.

As per ISO 4402, the method for calibrating APCs was based upon the size distribution of the silica-based A.C. Fine Test Dust (ACFTD). The size distribution was derived from the measurements using optical microscopes.

However, this method was replaced by another method as per the ISO standard 11171, as the supply of ACFTD was ceased in 1992. ISO Medium Test Dust (MTD) was selected as the replacement by the National Institute of Standards and Technology (NIST). This method uses a scanning electron microscope (SEM) with an image analysis software package to precisely identify the size and numbers of particles down to 1 μm.

Particle Size Classification
The particle size classification based on differential size ranges and cumulative sizes are given below:

SAE AS 4059 specifies the following differential size ranges of particles for the optical counting method, similar to that used in NAS standard: (1) 6 -14 μm(c), (2) 14 -21 μm(c), (3) 21 -38 μm(c), (4) 38 -70 μm(c), and (5) >70 μm(c).

SAE AS 4059 specifies the following cumulative sizes of particles for the automatic particle counting method using electron microscopes meant for new systems: (1) > 4 μm(c) (Code A), (2) > 6 μm(c) (Code B), (3) > 14 μm(c) (Code C), (4) > 21 μm(c) (Code D), (5) > 38 μm(c) (Code E), and (6) > 70 μm(c) (Code F).

Contamination Concentration Levels
SAE AS 4059 specifies the cleanness level of a given sample of fluid by a single figure representing the maximum allowed differential or cumulative particle counts (i.e. worst case), present in 100 ml of the fluid, for the designated particle sizes according to the particle counting method.

Cleanliness Classes for Differential Particle Counts
The cleanliness classes for the differential particle counts are given in Table A7.1.

Table A7.1 | Cleanliness Classes for Differential Particle Counts

Size		Maximum contamination limits, particles/100 ml				
		6 – 14 μm(c)	14 – 21 μm(c)	21 – 38 μm(c)	38 – 70 μm(c)	>70 μm(c)
Class	00	125	22	4	1	0
	0	250	44	8	2	0
	1	500	89	16	3	1
	2	1000	178	32	6	1
	3	2000	356	63	11	2
	4	4000	712	126	22	4
	5	8000	1425	253	45	8
	6	16000	2850	506	90	16
	7	32000	5700	1012	180	32
	8	64000	11400	2025	360	64
	9	128000	22800	4050	720	128
	10	256000	45600	8100	1440	256
	11	512000	91200	16200	2880	512
	12	1024000	182400	32400	5760	1024

Cleanliness Classes for Cumulative Particle Counts

The cleanliness classes for the cumulative particle counts are given in Table A7.2.

Table A7.2 | Cleanliness Classes for Cumulative Particle Counts

Size		Maximum contamination limits, particles/100 ml					
		>4 μm(c)	>6 μm(c)	>14 μm(c)	>21 μm(c)	>38 μm(c)	>70 μm(c)
		A	B	C	D	E	F
Class	000	195	76	14	3	1	0
	00	390	152	27	5	1	0
	0	780	304	54	10	2	0
	1	1560	609	109	20	4	1
	2	3120	1217	217	39	7	1
	3	6250	2432	432	76	13	2
	4	12500	4864	864	152	26	4
	5	25000	9731	1731	306	53	8
	6	50000	19462	3462	612	106	16
	7	100000	38924	6924	1224	212	32
	8	200000	77849	13849	2449	424	64
	9	400000	155698	27698	4898	848	128
	10	800000	311396	55396	9796	1696	256
	11	1600000	622792	110792	19592	3392	512
	12	3200000	1245584	221584	39184	6784	1024

Note: The information reproduced on Tables A7.1 and A7.2 is a brief extract from SAE AS4059. For further details and explanations refer to the full standard.

Appendix 8

Summary of Useful Relations and Definitions in Hydraulic Systems (in the English Units)

Volume (V)

$$V \text{ (ft}^3) = L \text{ (ft) x W (ft) x H (ft)}$$
$$1 \text{ ft}^3 = 1728 \text{ in}^3$$

Density (ϱ)

$$\text{Density, } \varrho \text{ (Slug/ft}^3) = \text{Mass (Slug) / Volume (ft}^3)$$

Specific Gravity (SG)

Specific gravity (SG) = Density of an object / Density of water

Force (F)

$$F \text{ (lb)} = M \text{ (Slug) x a (ft/s}^2)$$

Work (W)

$$W \text{ (in.lb)} = F \text{ (lb) x d (in)}$$
$$W \text{ (ft.lb)} = F \text{ (lb) x d (ft)}$$

Power

$$\text{Power (ft.lb/s)} = F \text{ (lb) x v (ft/s)}$$

Horse Power (HP)

$$HP = \frac{F \text{ (lb) x v (ft/s)}}{550}$$

Torque (T)

$$T \text{ (ft.lb)} = F \text{ (lb)} \times r \text{ (ft)}$$

Torque – Power Relations

$$\text{Power (hp)} = T \text{ (ft.lb)} \times \omega \text{ (rad/s)} / 550$$
$$\text{Power (hp)} = T \text{ (ft.lb)} \times N \text{ (rpm)} / 5252$$
$$\text{Power (hp)} = T \text{ (in.lb)} \times N \text{ (rpm)} / 63025$$

Energy

$$\text{Energy, } E \text{ (ft.lb)} = F \text{ (lb)} \times d \text{ (ft)}$$

Hydraulic pressure

$$P \text{ (psi)} = F \text{ (lb)} / A \text{ (in}^2)$$
$$1 \text{ bar} = 14.5 \text{ psi}$$

Hydraulic Force

$$F \text{(lb)} = P \text{(psi)} \times A \text{(in}^2)$$

Absolute Viscosity (µ)

The absolute viscosity of a fluid bounded between a stationary plate and a thin movable plate of surface area 'a' located at a distance 'd' from the stationary plate when the movable plate is subjected to a force 'F' and moves with the velocity 'v' is given by:

$$\text{Absolute viscosity, } \mu = \frac{(F/a)}{(v/d)}$$

Units of Absolute Viscosity

1 Poise = 1 dyne second per square centimetre (1 dyne.s/cm^2)
1 centipoise (cP) = 0.01 Poise
1 Pascal second (Pa.s) = 1 N.s/m^2 = 0.02088543 lb.s/ft^2
1 Poise = 0.1 Pa.s

Kinematic viscosity (ν)
Kinematic viscosity is the measure of fluid's resistance to flow under gravity and is given by the absolute viscosity (μ) divided by the fluid density (ϱ), at a given temperature.

$$\text{Kinematic viscosity, } \nu = \frac{\mu}{\varrho}$$

Units of Kinematic Viscosity

$$1 \text{ Stoke} = 1 \text{ cm}^2/\text{s}$$
$$1 \text{ Centi Stoke (cSt)} = 0.01 \text{ Stoke} = 1 \text{ mm}^2/\text{s}$$
$$1 \text{ cSt} = 0.00155 \text{ in}^2/\text{s} = 0.00001076391 \text{ ft}^2/\text{s}$$

Saybolt Universal Seconds (SUS)
It is the time measured in seconds required for 60 ml of a sample of test fluid to pass through the calibrated orifice of a Saybolt Universal viscometer at the specified temperature, as covered by the empirical test method specified in the standard ASTM D88.

$$\nu \text{ (cSt)} = \mu \text{ (cP)} / \text{Specific Gravity}$$
$$\nu \text{ (SSU)} = 4.63 \, \mu \text{ (cP)} / \text{SG}$$
$$\text{cSt} = \text{SSU}/4.635$$

Viscosity Classification Systems
Some of the ISO Viscosity Grades, as per the ISO standard 3448:1992, are as follows: 2, 3, 5, 7, 10, 15, 22, 32, 46, 68, 100, 150, 220, 320, 460, 680, 1000, 1500 and more.

Viscosity Index (VI)
Hydraulic fluids can typically be selected with values of viscosity index (VI) in the range from 90 to 110.

Flow Rate, (Q)
The volumetric flow rate is a measure of the volume of the fluid passing a given cross-sectional area per unit of time. It is measured in ft^3/s, gpm, etc.

Flow Velocity (v)

It is the average velocity of the molecules in the moving fluid in a hydraulic system.

Flow Rate Vs Velocity of Flow

The flow rate of a fluid pumped through a pipeline having a cross-sectional area A and travelling with a velocity v is given by:

Flow rate (Q, ft³/s) = Area (A, ft²) × Velocity (v, ft/s)

Reynolds Number (R_e)

$$R_e = \frac{v\,D\,\varrho}{\mu} = \frac{v\,D}{\upsilon}$$

v = Fluid velocity, ft/s | D = Internal diameter of the pipe, ft | ϱ = Fluid density, slug/ft³ | μ = Absolute viscosity of the fluid, lb·s/ft² | υ (nu) = Kinematic viscosity, ft²/s

Using the kinematic viscosity in cSt, the Reynolds number is given by:

$$\text{Reynolds number, } Re = \frac{7740 \times v(\text{ft/s}) \times d(\text{in})}{\nu\,(\text{cSt})}$$

Re, critical ~2000
For Re < 2000, the flow is laminar
For Re > 2000, the flow is turbulent

Fluid Compressibility and Bulk Modulus

Bulk modulus, B = 1/Compressibility
Bulk modulus, B (psi) = - ΔP (psi)/ (ΔV/V)

ΔP =Differential change in the pressure, in psi
ΔV =Differential change in the fluid volume, in in³
V =The original volume of the fluid, in in³

Categories of Hydraulic Fluids

- Mineral-based (Petroleum oil)
- Fire-resistant (Water-based and synthetic)
- Bio-degradable
- Food grade

Classification of Fire-resistant Hydraulic Fluids
(As per ISO 6743-4 /CETOP RP 77H)

- **HFA:** Oil-in-water emulsions with a combustible proportion of 20% maximum.
- **HFB:** Water-in-oil emulsions with a combustible proportion of 60% maximum.
- **HFC:** Water glycol solutions with a water proportion of at least 35%.
- **HFD:** Water-free fluids on a synthetic base.

NAS Code
The National Aerospace Standard (NAS) 1638 coding system defines the maximum numbers permitted of 100 ml volume at various size intervals (for aircraft systems)

SAE Aerospace Standard AS4059
The SAE aerospace standard AS4059, for specifying particulate contamination in hydraulic fluids in different classes, was developed in 1988 as a replacement to the NAS 1638 standard.

Mesh Number/Sieve Number, Wire-mesh Filter
It is the number of openings from the centre of any wire of the wire mesh to the centre of the parallel wire one inch away.

Beta Ratio, Filter

$$\text{Beta ratio}_{x(c)} = \frac{\text{Particle count in the upstream fluid}}{\text{Particle count in the downstream fluid}}$$

Filter Efficiency, Filter

$$\text{Efficiency}_{(x)} = \left(1 - \frac{1}{\beta}\right) \times 100$$

Absolute Micron Rating, Filter
It is the smallest size of particles a filter can capture in excess of 98.6% on the first pass through it.

Nominal Micron Rating, Filter
It is the smallest size of particles a filter can capture in a specified quantity, in the range from 50 to 95% on the first pass through it.

Differential Pressure (ΔP), Filter
The differential pressure (ΔP) across a filter indicates the difference between its inlet and outlet pressures when a fluid flows through it.

Particle Capture Efficiency (Dirt Holding Capacity), Filter
It indicates the quantity of the solid dirt that a filter element can hold before it has to be replaced.

Burst (Collapse) Pressure, Filter
It is the minimum inside-out (or outside-in) pressure differential that a filter can withstand without the outward structural or media failure.

Sizing of Hydraulic Reservoir

Reservoir size, gallons = (3 to 5) x pump flow rate, gpm

Heat Load of a New Reservoir

Heat Load of a new reservoir = 30 to 50% of the motor rating

Heat Load of an Existing Reservoir

$$\text{Heat load (hp)} = \frac{V \text{ (gallons) x } Cp \text{ (btu/lb) x } \varrho \text{ (lb/ft}^3\text{) x } \Delta T \text{ (°F)}}{317.3 \text{ x } \Delta t \text{ (min)}}$$

V = Tank volume | ΔT = Temperature difference during system operation from start to some duration (Δt) | Cp is the specific heat of the fluid in btu/lb (°F) | ϱ is the density of the fluid in lb/ft^3

$$1 \text{ hp} = 2544 \text{ btu/hour}$$

Heat dissipation by Hydraulic Reservoirs

$$H \text{ (hp)} = 0.001 \text{ x } \Delta T \text{ (°F) x } A \text{ (ft}^2\text{)}$$

Pressure Rating, Pump
It is the maximum pressure that a pump can be subjected to without the risk of its pressure-related failures.

Volumetric Displacement (V_D), Pump
It is the volume of the fluid that is carried by a pump in one revolution of its driveshaft. Units: in^3/rev

Theoretical Flow Rate (Q_T), Pump

$$Q_T \text{ (in}^3\text{/min)} = V_D \text{ (in}^3\text{/rev) x } N \text{ (rpm)}$$
$$Q_T \text{ (gpm)} = Q_T \text{ (in}^3\text{/min) / 231}$$

Pump Slippage (Q_s)

$$\text{Pump slippage, } Q_S = \text{Internal fluid leakage}$$

Actual Flow Rate (Q_A), Pump

$$\text{Actual flow rate, } Q_A = \text{Theoretical flow rate, } Q_T - \text{Slippage, } Q_S$$

Actual Torque (T_A), Pump

$$T_A(\text{in.lb}) = \frac{\text{Actual power delivered to the pump (hp) x } 63025}{N \text{ (rpm)}}$$

$$T_A(\text{ft.lb}) = \frac{\text{Actual power delivered to the pump (hp) x } 5252}{N \text{ (rpm)}}$$

$$T_A(\text{ft.lb}) = \frac{\text{Actual power delivered to the pump (hp) x } 550}{n \text{ (rps)}}$$

Theoretical Torque (T_T), Pump

$$\text{Theoretical Torque, } T_T(\text{in.lb}) = \frac{V_D(\text{in}^3/\text{rev}) \text{ x P (psi)}}{2\Pi}$$

Input Power, Pump

$$\text{Pump input power (hp)} = \frac{T_A(\text{in.lb}) \text{ x N(rpm)}}{63025}$$

$$\text{Pump input power (hp)} = \frac{T_A \text{ (ft.lb) x N(rpm)}}{5252}$$

$$\text{Pump input power (hp)} = \frac{T_A \text{ (ft.lb) x n(rps)}}{550}$$

Output Power, Pump

$$\text{Pump output power (hp)} = \frac{P(\text{psi}) \text{ x } Q_A(\text{gpm})}{1714}$$

Volumetric Efficiency (η_v), Pump

$$\text{Volumetric Efficiency } (\eta_v) = \frac{\text{Actual flow rate}}{\text{Theoretical flow rate}} = \frac{Q_A}{Q_T}$$

Mechanical Efficiency (η_m), Pump

$$\text{Mechanical Efficiency} \left(\eta_m\right) = \frac{\text{Pump output power, assuming no leakage}}{\text{Actual power delivered to the pump}}$$

$$\text{Mechanical Efficiency} \left(\eta_m\right) = \frac{P \times Q_T}{T_A \times N}$$

$$\text{Mechanical Efficiency} \left(\eta_m\right) = \frac{T_T}{T_A}$$

Overall Efficiency (η_o), Pump

$$\text{Overall Efficiency} \left(\eta_o\right) = \frac{\text{Actual power delivered by the pump}}{\text{Actual power delivered to the pump}}$$

$$\text{Overall Efficiency} \left(\eta_o\right) = \frac{P \times Q_A}{T_A \times N}$$

$$\eta_o = \eta_v \times \eta_m$$

Maximum operating pressure (P), Cylinder
It is the maximum pressure that a cylinder can be subjected to without the risk of its pressure-related failures.

Bore Diameter (D), Cylinder
It is the diameter at a cylinder bore.

Piston-rod Diameter (d), Cylinder
It is the diameter of a cylinder piston-rod.

Stroke Length (L), Cylinder
It is the linear movement that a cylinder can produce.

Maximum Stroke Length, Cylinder

It is the maximum linear movement that a cylinder can produce.

Thrust/Pull (F), Cylinder

Thrust, F (lb) $= P$ (psi) x A_{ext} (in²)

Pull, F (lb) $= P$ (psi) x A_{ret} (in²)

A_p is the piston area | A_r is the piston-rod area | A_{ext} is the active area during extension: $(A_{ext} = A_p)$ | A_{ret} is the active area during retraction: $(A_{ret} = A_p - A_r)$

Input Power, Cylinder

$$P_{input} \text{ (hp)} = P \text{ (psi)} \times Q_A \text{ (gpm)}/1714$$

Output Power, Cylinder

$$P_{output} \text{ (hp)} = \text{Force (lb)} \times \text{Velocity (ft/s)} /550$$

Speed Relation, Cylinder

$$Q \text{ (ft}^3/\text{s)} = A \text{ (ft}^2) \times v \text{ (ft/s)}$$

Operating Pressure (P), Hydraulic Motor

It is the maximum pressure that a hydraulic motor can be subjected to without the risk of its pressure-related failures.

Displacement (V_D), Hydraulic Motor

It is the volume of fluid required for turning the output shaft of a motor through one revolution. Unit: in³/rev.

Theoretical Flow Rate (Q_T), Hydraulic Motor

$$Q_T \text{ (gpm)} = \frac{V_D (\text{in}^3/\text{rev}) \times N(\text{rpm})}{231}$$

Slippage, Hydraulic Motor
It is the internal leakage of fluid through the unintended paths of the motor, without performing any useful work.

Speed, Hydraulic Motor

$$\text{Speed, N (rpm)} = \frac{\text{Theoretical motor flow, } Q_T \text{ (gpm) x 231}}{\text{Motor displacement, } V_D \text{ (in}^3/\text{rev)}}$$

Maximum speed, Hydraulic Motor
It is the speed of the motor, at a particular inlet pressure, that it can sustain for a limited period without damage to the motor.

Minimum motor speed, Hydraulic Motor
It is the slowest, continuous, rotational speed obtainable from the output shaft of the motor.

Input Power (P_{in}), Hydraulic Motor

$$\text{Input Horse Power (hp)} = \frac{P\text{(psi) x } Q_A\text{(gpm)}}{1714}$$

Theoretical Torque (T_T), Hydraulic Motor

$$\text{Theoretical Torque, } T_T\text{(in.lb)} = \frac{V_D\text{(in}^3/\text{rev) x } \Delta P\text{(psi)}}{2\Pi}$$

Breakaway (Starting) Torque, Hydraulic Motor
It is the rotary force required for turning a stationary load connected to a motor.

Running Torque, Hydraulic Motor
It is the torque required to run a load connected to a motor.

Stalling Torque, Hydraulic Motor
It is the torque needed to stop a motor to a standstill.

Actual Torque (T_A), Hydraulic Motor

It is the torque which a motor develops to drive the attached load alone.

Output Power (P_{out}), Hydraulic Motor

$$\text{Output Horse Power (hp)} = \frac{T_A (\text{in.lb}) \times N (\text{rpm})}{63025}$$

Volumetric Efficiency (η_v), Hydraulic Motor

$$\text{Volumetric efficiency, } (\eta_v) = \frac{\text{Theoretical flow rate } (Q_T)}{\text{Actual flow rate } (Q_A)}$$

Mechanical efficiency (η_m), Hydraulic Motor

$$\text{Mechanical efficiency, } (\eta_m) = \frac{\text{Actual torque, } (T_A)}{\text{Theoretical torque } (T_T)}$$

Overall Efficiency (η_o), Hydraulic Motor

$$\text{Overall efficiency, } (\eta_o) = \frac{\text{Brake power delivered by the motor}}{\text{Hydraulic power delivered to the motor}}$$

$$= \eta_v \times \eta_m$$

Flow coefficient (Kv), Hydraulic Valve

It is the flow rate (m^3/h) of water flowing through a valve at a temperature in the range from 5 to 30°C that causes one bar pressure drop across it.

Flow coefficient (Cv), Hydraulic Valve

It is defined as the flow rate (gpm) of the water flowing through a valve at the temperature of 60°F that causes one psi pressure drop across it.

Flow Rate (in General), Hydraulic Valve

$$Q = Cv \times \sqrt{(\Delta P / SG)}$$

Q is the flow rate in gpm | ΔP is the pressure drop across the valve in psi | Cv is the flow coefficient in 'gpm/$\sqrt{psi}$' | SG denotes the specific gravity

Accumulator Volume for Charging/Discharging under Isothermal Condition

Accumulator volume, $V_0 = [V_1 - V_2] / [(P_0 / P_1) - (P_0 / P_2)]$

Volume for the Full Cylinder Extension

Volume for full cylinder extension, $(\Pi D^2 / 4) \times S = V_1 - V_2$

Accumulator Volume for Charging/Discharging under Adiabatic Condition

$$V_{0,adiabatic} = [V_1 - V_2] / [(P_0 / P_1)^{1/n} - (P_0 / P_2)^{1/n}]$$

Accumulator Volume for Slow Charging and Quick Discharging

$$V_0 = [V_1 - V_2] / \{(P_0 / P_2)^{1/n} - [(P_2 / P_1)^{1/n} - 1]\}$$

Accumulator Volume under Temperature Influence

$$V_{oT} = V_0 \times (T_2/T_1)$$

Accumulator Volume at Higher Pressures

$$V_{oP} = V_0 / C_i \text{ (for isothermal condition)}$$
$$V_{oP} = V_0 / C_a \text{ (for adiabatic condition)}$$
C_i - Correction coefficient for isothermal condition
C_a - Correction coefficient for adiabatic condition

Diametrical Size, Fluid Conductor
The diametrical size of a conductor is specified by its inside diameter, outside diameter, or nominal size.

Inside Diameter, D_i, Fluid Conductor
It is the smallest cross-sectional diameter of a conductor.

Outside Diameter, D_o, Fluid Conductor
It is the largest cross-sectional diameter of a conductor.

Nominal Size, Fluid Conductor
In ANSI and SAE standards, for example, the size of a pipe is specified in terms of nominal pipe size (NPS), and in SI system, it is specified in terms of Nominal Diameter (DN).

Wall Thickness (t), Fluid Conductor

$$\text{Wall thickness, } t = (D_o - D_i) / 2$$

Schedule Number, Fluid Conductor
In ANSI/SAE system, the wall thickness of a pipe is described in terms of a 'schedule number'. The schedule numbers vary from 5 through 160. Altogether, there are eleven different schedule numbers. They are: 5, 10, 20, 30, 40, 60, 80, 100, 120, 140, and 160.

Hoop Stress, Fluid Conductor

$$\text{Hoop stress} = P \times D_i / 2t$$

Burst Pressure, Fluid Conductor

$$\text{Burst pressure (BP)} = 2tS / D_i$$

Working Pressure, Fluid Conductor

$$\text{Working pressure (WP)} = \frac{\text{Burst pressure (BP)}}{\text{Safety factor (SF)}}$$

Design Pressure, Fluid Conductor
It is the pressure to which each component of a piping system is designed.

Maximum Allowable Working Pressure,
It is the maximum pressure of a piping system, determined by the weakest component of a piping system. It is not to exceed its design pressure.

Minimum Bend Radius, Fluid Conductor
It is the smallest radius of the curved section of a conductor (tube or hose) beyond which it should not be bent without flattening, kinking or wrinkling.

Design Temperature
It is the maximum temperature at which a piping component is designed to operate.

Fluid Velocity

$$Q \ (ft^3/s) = A(ft^2) \ x \ v(ft/s)$$

$$v(ft/s) = \frac{0.3208 \ x \ Q \ (gpm)}{A \ (in^2)}$$

Fluid Velocities – Suction Lines

Viscosity (cSt)	Maximum velocity (ft/s)
150	2
100	2.5
50	3.6
30	4

Fluid Velocities - Pressure Lines

Pressure line	For flow rate >2.6 gpm
900 – 1450 psi	13 – 14
1450 – 2300 psi	14 – 16
2300 – 3600 psi	16 – 18
3600 – 5800 psi	18 – 20

Fluid Velocities – Return Lines

Fluid velocities to be utilized for initial pipe sizing in return lines should be between 6.5 to 10 ft/s.

Dimensioning Based on Flow Velocity

When using the dimensioning method based on flow velocity, the inner diameter of the pipe can be determined by using the equation below, when a maximum flow rate and recommended flow velocity are known.

$$d = \sqrt{\frac{4 \times Qmax}{\Pi \times v}}$$

d =Inner diameter of the pipe (ft) | Q_{max} =Maximum flow rate (ft^3/s) | v =Flow velocity (ft/s)

Dimensioning based on Pressure Losses

The overall pressure losses in piping include frictional pressure losses arising in straight pipe sections, as well as individual pressure losses in bends and junctions.

When using the dimensioning method based on the pressure losses, the inner diameter of a pipe is selected so that the resulting pressure losses do not increase above a specified value. The total pressure loss is allowed to be 3 to 5% for systems in continuous use. The total pressure loss is allowed to be 7 to 10% for systems with intermittent duty cycle.

Frictional Pressure Losses in Pipes and Hoses

$$\Delta Pa = \lambda \cdot \frac{l}{d} \cdot \frac{\varrho\, v^2}{2}$$

Δp_a=Frictional pressure loss, [lb/ft^2], λ=Frictional resistance factor, l = Length of the pipe [ft], d = pipe id [ft], ϱ = Hydraulic fluid density [slug], v = Flow velocity [ft/s]

$$[1 \text{ lb/ft}^2 = 1/144 = 0.00694444 \text{ psi}]$$

Friction Factor for Laminar Flow

$$\text{Friction factor, } \lambda = \frac{64}{Re}$$

Frictional Pressure Losses in Straight Pipe Sections

$$\text{Pressure loss, } \Delta Pa = \frac{32\,\mu\, l\, v}{d^2}$$

$$\text{Pressure loss, } \Delta Pa = \frac{128\,\mu\, l\, Q}{\pi\, d^4}$$

Δp_a=Frictional pressure loss, [lb/ft^2], μ = Absolute viscosity (lb.s/ft^2), l = Length of the pipe [ft], d = pipe id [ft], v = Flow velocity [ft/s], Q = Flow rate (ft^3/s)

Relative Roughness

$$\text{Relative roughness} = \varepsilon/d$$

Typical values of absolute roughness: Drawn tubing – 0.0015 mm, Cast iron – 0.26 mm, Riveted steel – 1.8 mm

Friction Factor in Smooth Pipes, Turbulent flow

$$\text{Friction factor, } \lambda = \frac{0.316}{\text{Re}^{0.25}}$$

Frictional Losses in Smooth Pipes, Turbulent flow

$$\Delta Pa = 0.214 \frac{\mu^{0.25} \, l \, \varrho^{0.75} \, Q^{1.75}}{d^{4.75}}$$

Friction Factor in Rough Pipes, Turbulent Flow

$$\text{Friction factor, } \lambda = \frac{0.25}{[\log_{10}(\varepsilon/3.7 \, d) + \left(5.74/\text{Re}^{0.25}\right)]^2}$$

Individual Pressure Losses

$$\Delta Pb = \zeta \frac{\varrho \cdot v^2}{2} = \zeta \frac{\varrho \cdot Q^2}{2 \, A^2}$$

Δp_b = individual pressure loss [lb/ft²], ζ = individual resistance factor, ϱ = the fluid density [slug], v = flow velocity [ft/s]

Values of Loss Coefficient (ζ)

90⁰ elbow – 0.2, 45⁰ elbow – 0.15, Tee fitting – 0.9, Sharp-edged entrance – 0.5, Rounded entrance – 0.05, Sharp-edged exit – 1.0, Rounded exit – 1.0

References

1. Aerospace Fluid Power - Contamination Classification for Hydraulic Fluids AS4059F
   https://www.sae.org/standards/content/as4059f/
2. Article on 'Hydraulic Circuit Design & Analysis', Dr. Sunil Jha.
3. Article on" 'Computerized Design Analysis of Machine Tool Hydraulic System Dynamics', Dr. Ing T. Hong, P.E.and Dr. Richard K. Tessmann, P.E., An FES/BarDyne Technology Transfer Publication.
4. Automatic particle counters for fluid contamination control by Noria Corporation
5. Document on 'BRITISH FLUID POWER ASSOCIATION, QUALIFICATIONS HYDRAULIC SYSTEM DESIGN PROGRAMME (HSD4) CETOP (PASSPORT) Issue 2, 2004OCCUPATIONAL LEVEL 4'.
6. Fluid Condition Handbook, SYSTEM CONDITIONS, MP FILTRI S.p.A. www.mpfiltri.com
7. Lubricating Oil Laboratory (Swansea), Swansea
8. Particle Measurement Technology in Practice. From Theory to Application, HYDAC Filtertechnik GmbH, Industriegebiet 66280 Sulzbach / Saar, Germany, www.hydac.com
9. Swift-JB International, LLC is a division of Swift Filters, Inc.

Fluid Power Educational Series Books

1. Pneumatic Systems and Circuits -Basic Level (In the SI Units)
2. Industrial Pneumatics -Basic Level (In the English Units)
3. Pneumatic Systems and Circuits -Advanced Level
4. Electro-Pneumatics and Automation
5. Design of Pneumatic Systems (In the SI Units)
6. Design Concepts in Pneumatic Systems (In the English Units)
7. Maintenance, Troubleshooting, and Safety in Pneumatic Systems
8. Industrial Hydraulic Systems and Circuits -Basic Level (In the SI Units)
9. Industrial Hydraulics -Basic Level (In the English Units)
10. Hydraulic Fluids
11. Hydraulic Filters: Construction, Installation Locations, and Specifications
12. Hydraulic Power Packs (In the SI Units)
13. Power Packs in Hydraulic Systems (In the English Units)
14. Hydraulic Cylinders (In the SI Units)
15. Hydraulic Linear Actuators (In the English Units)
16. Hydraulic Motors (In the SI Units)
17. Hydraulic Rotary Actuators (In the English Units)
18. Hydraulic Accumulators and Circuits (In the SI Units)
19. Accumulators in Hydraulic Systems (In the English Units)
20. Hydraulic Pipes, Tubes, and Hoses (In the SI Units)
21. Pipes, Tubes, and Hoses in Hydraulic Systems (In the English Units)
22. Design of Industrial Hydraulic Systems (In the SI Units)
23. Design Concepts in Industrial Hydraulic Systems (In the English Units)
24. Maintenance, Troubleshooting, and Safety in Hydraulic Systems
25. Hydrostatic Transmissions (HSTs) (In the SI Units)
26. Concepts of Hydrostatic Transmissions (In the English Units)
27. Load Sensing Hydraulic Systems (In the SI Units)
28. Concepts of Load Sensing Hydraulic Systems (In the English Units)
29. Electro-hydraulic Proportional Valves
30. Electro-hydraulic Servo Valves
31. Cartridge Valves
32. Electro-hydraulic Systems and Relay Circuits
33. Practical Book: Pneumatics - Basic Level
34. Practical Book: Electro-pneumatics - Basic Level
35. Practical Book: Industrial Hydraulics – Basic Level
36. Programmable Logic Controllers and Programming Concepts

For more details, please visit: **https://jojibooks.com**

About the Author

Joji Parambath is a trainer in the field of Pneumatics, Hydraulics, and PLC, for over 25 years. During his career, he has trained numerous professionals from the industries as well as faculty members and students of engineering institutions.

At present, he is the key trainer at Fluidsys Training Centre, Bangalore, India, (https://fluidsys.org) which is providing training in the field of Pneumatics and Hydraulics. He has already written two books on Pneumatics and Hydraulics. The publication of the present series of 36 books is intended to restructure and update the existing books.

The author wishes to thank all trainees for their lively interaction and many useful suggestions during the training programmes that prompted the author to write the present series of books. You may send your feedback to joji.p@hotmail.com

10th June 2020